D0374286

ELEMENTARY
STATISTICS
FOR
ECONOMICS
AND
BUSINESS

Selected Readings

ELEMENTARY STATISTICS FOR ECONOMICS AND BUSINESS

Selected Readings

edited by

EDWIN MANSFIELD

Wharton School
University of Pennsylvania

W·W·NORTON & COMPANY·INC·

NEW YORK

"Living with Symbols" by Arthur M. Ross: from *American Statistician* (June 1966). Reprinted by permission of the author and the American Statistical Association.

"Sources and Errors of Economic Statistics": from *On the Accuracy of Economic Observations* by Oskar Morgenstern (Princeton University Press, 1963). Reprinted by permission of the publishers.

"How to Lie with Statistics": from *How to Lie with Statistics* by Darrell Huff. Pictures by Irving Geis (W. W. Norton & Company, Inc., 1954). Reprinted by permission of the publishers.

"Probability: An Objectivist View": from *Probability, Statistics, and Truth* by Richard von Mises (New York: Humanities Press, Inc., 1939). Reprinted by permission of the publishers.

"Statistical Inference": from *The Design of Experiments* (8th edition) by R. A. Fisher (Oliver and Boyd Limited, 1966). Reprinted by permission of the Literary Executor of the late Sir Ronald A. Fisher and Oliver and Boyd Limited, Edinburgh.

"The Bayesian Approach to Statistical Decision" by J. Hirshleifer: from the *Journal of Business* (October 1961). Reprinted by permission of the University of Chicago Press.

"Probability: A Subjectivist View": from *The Foundations of Statistics* by L. J. Savage (John Wiley & Sons, Inc., 1954). Reprinted by permission of the publishers.

"Dependable Samples for Market Surveys" by Morris H. Hansen and William N. Hurwitz: from the *Journal of Marketing* (October 1949). Reprinted by permission of the American Marketing Association.

"Some Applications of Statistics for Auditing" by John Neter: from the *Journal of the American Statistical Association* (March 1952). Reprinted by permission of the author and the American Statistical Association.

"Tests of Hypotheses Regarding Bilateral Monopoly": from *Bargaining and Group Decision Making* by L. Fouraker and S. Siegel (McGraw-Hill Book Company, 1960). Used with permission of McGraw-Hill Book Company.

"Gibrat's Law and the Growth of Firms" by Edwin Mansfield: from the *American Economic Review* (1962). Reprinted by permission of the American Economic Association.

"Sales Forecasting by Correlation Techniques" by Robert Ferber: from the *Journal of Marketing* (January 1954). Reprinted by permission of the author and the American Marketing Association.

"What do Statistical 'Demand Curves' Show?" by E. J. Working: from the *Quarterly Journal of Economics* (1927). Reprinted by permission of Harvard University Press.

"Statistical Cost Functions of a Hosiery Mill" by Joel Dean: from the *Journal of Business* (1941). Reprinted by permission of the University of Chicago Press.

"Economies of Scale: Some Statistical Evidence" by Frederick T. Moore: from the *Quarterly Journal of Economics* (May 1959). Reprinted by permission of Harvard University Press.

"Great Ratios of Economics" by Lawrence R. Klein and Richard F. Kosobud: from the *Quarterly Journal of Economics* (May 1961). Reprinted by permission of Harvard University Press.

"The Relation between Unemployment and the Rate of Change of Money Wage Rates in the United Kingdom, 1861-1957" by A. W. Phillips: from *Economica* (November 1958). Reprinted by permission of the London School of Economics and Political Science.

"Fluctuations in the Savings-Income Ratio: A Problem in Economic Forecasting" by Franco Modigliani: from *Studies in Income and Wealth*, Vol. 11 (1949). Reprinted by permission of the National Bureau of Economic Research, Inc.

*To Sally, the Best Mathematician
and Most Dedicated Teacher I Know*

Contents

Preface

The study of statistics is at the heart of a modern education in economics and business. Statistics provides tools and techniques for both research and practical decision-making. The popular conception of statistics as the collection and presentation of large masses of data touches on only a minor part of the field. Modern courses in statistics are concerned with the ways in which one can derive valid conclusions from empirical evidence. The emphasis is on analysis, not simple description. For example, statistical techniques are used to indicate the extent to which an estimate of the elasticity of demand for a particular product may be in error, and how this error can be reduced. Statistics, as one distinguished statistician puts it, is "the technology of the scientific method." [1] It is also the study of decision-making under uncertainty. For example, statistics helps to provide the decision-maker with decision rules designed to recognize the uncertainties of the situation and to further his objectives, whatever they may be.

The purpose of this book is to provide supplementary readings for elementary courses in economic and business statistics. The book illustrates the use of the techniques that are presented in these courses, introduces the student to some well-known articles based on the utilization of these techniques, and permits him to read some of the classic statements regarding controversial issues at the foundations of statistics. The purpose is not to provide a substitute for a statistics textbook. There are many excellent statistics texts available for elementary courses. The aim of this book is to supplement them by exposing the student in greater depth to the techniques discussed in the textbooks.

The need for a book of this sort is obvious, I think, to many teachers. The available textbooks concentrate on the description and derivation of the standard statistical tools, and seldom present examples of their application to real and important problems. To whet

[1] Alexander Mood, *Introduction to the Theory of Statistics*, McGraw-Hill, 1950, p. 1.

the interest of students and to deepen their understanding, it is necessary to use supplementary readings to illustrate the application of these techniques; but no book of this sort exists at present. Hopefully, this volume will help to fill the gap. Each of the articles has been chosen with an eye to the needs and capacities of the typical student in elementary courses in economics and business statistics. An effort has been made to include articles that provide important substantive results, as well as illustrate the use of basic techniques.

Part I introduces the student to the sources and limitations of economic statistics as well as common errors in their interpretation. Part II is concerned with probability and statistical inference, and Part III contains applications of sampling theory and of the t, χ^2, and F tests. Simple regression is taken up in Parts IV and V, Part IV being concerned with demand and costs, and Part V dealing with income, employment, and consumption.

This book has developed from the statistics courses that I have taught over the past decade at the Graduate School of Industrial Administration at Carnegie Institute of Technology (now Carnegie-Mellon University) and at the Wharton School of the University of Pennsylvania. My students, by their reactions, have contributed significantly to the choice of papers. Also, my thanks go to Professors Roger Bolton of Williams College, Gerald Eyrich of Claremont Men's College, Marc Nerlove of Yale University, and Richard N. Rosett of the University of Rochester for their helpful comments on the manuscript.

E. M.

Philadelphia
May 1969

ELEMENTARY
STATISTICS
FOR
ECONOMICS
AND
BUSINESS

Selected Readings

ECONOMIC STATISTICS: NATURE, LIMITATIONS, AND PITFALLS

1

To be a reasonably effective and sophisticated member of his profession, an economist or a manager of a business must be able to work with economic statistics, which, after all, are the empirical bases for these professions. The first step in learning to work with economic data is to consider their nature, derivation, and limitations. Economic statistics differ in many important respects from the data that arise in the physical sciences. They have special limitations and pitfalls that the student should understand. The purpose of the four articles in Part One is to explain and illustrate the characteristics of economic data.

The opening article, by Arthur Ross, emphasizes that, in the economic and social sphere, "statistical truths . . . are created rather than discovered." The statistician invents and defines the categories that are used, the dimensions that are measured, and the way in which these dimensions are used to characterize complex social conditions and relationships. For example, the concept of "poverty" is subjective to a considerable extent, and consequently the measures of poverty put forth by the economic statistician are incomplete. Ross is careful to point out that he is "not suggesting that statisticians should be blamed for failing to measure the subjective, social, and spiritual aspects of [reality]. But perhaps they do have some responsibility to warn the layman against the danger of confusing the shadow with the substance."

In the next article, Oskar Morgenstern points out and discusses the various kinds of errors that arise in economic statistics. He stresses that, unlike physical scientists, economists cannot as a rule derive data through designed experiments, and it is seldom feasible for data users to be aware of the detailed nature of the data's derivation. Moreover, economic and social statistics are sometimes "based on evasive answers and deliberate lies of various types"; they "are frequently not·gathered by highly trained observers but by personnel gathered *ad hoc*"; and they are frequently derived from faulty questionnaires. The user of economic statistics must constantly be on his guard: data containing large errors of this sort can be worse than no data at all.

Even if the data contain no such errors, they can be used to create false impressions. In the following article, Darrell Huff catalogues a number of common ways that one can, intentionally or unintentionally, "lie with statistics." He points out that averages can sometimes be quite misleading: different types of averages can give quite different results and the variation about an average can be very important. He also shows how small samples, inadequate controls, and ambiguous graphs can mislead the unwary. Although Huff presents his material in a light and entertaining manner, the plight of the person who falls prey to these misleading statistics can be very serious indeed.

Finally, the article by the Bureau of Labor Statistics describes one of the best-known and most important economic statistics: the Consumer Price Index. This index is used widely by the general public, by the economics profession, and in labor-management contracts to adjust wages. This article describes how the index is constructed, the way in which the basic data are collected, the history of the index, and its uses. It also discusses some of the limitations of the index, part of the article by Ross also having touched on this topic. This article illustrates the construction and use of index numbers, which are an important branch of economic statistics.

Living
with
Symbols

ARTHUR M. ROSS

Arthur Ross is Professor of Economics at the University of
Michigan and a former United States Commissioner of
Labor Statistics. This article appeared in the *American
Statistician* in 1966.

Let us recognize candidly that statistical truths, like
the other truths about man's social life, are created rather than
discovered. It may well be different when it comes to measuring the amount of rainfall or the population of redwood trees.
These are physical phenomena. But when it comes to unemployment or poverty or price inflation or mental disease, we are
dealing with social phenomena. It is man who invents and defines
these categories. It is man who selects a few dimensions that are
capable of measurement and uses them to characterize complex
social conditions and relationships. It is man who decides how
much effort should be expended in measuring these dimensions
or others that might be selected.

These facts are so poorly understood that many consequent
misunderstandings result. As an example we may take the problem of defining full employment. The press and the public appear to believe that full employment has objective reality like a
tree or an inch of rain. Is it 4 percent, or 3 percent, or 3.5? they
ask impatiently. Why can't all you experts agree? The fact is that
full employment is not a statistical concept but a policy problem.
The full employment rate is reached when the nation decides

that the costs of reducing unemployment even further are greater than the benefits. Obviously much depends on the value system of those who are making the decision on behalf of the nation. Much depends on what has been done to reduce the costs of abundant work opportunity through manpower programs, anti-inflation measures, improvements in international finance, etc. Much depends on price and wage trends in other countries. My own belief is that we won't know the full employment rate until we have almost reached it. When the unemployment rate was 7 percent, it was rather fruitless to debate whether the constraints would be overwhelming at 4 percent. If the rate falls to the neighborhood of 3.5 percent, then it will be time to debate whether it is practical to reduce it to 3.0 or 2.5 or not at all.

Equally subjective is the concept of poverty. I will not deny the operational usefulness of some dividing line such as a $3000 family income, especially after it has been adjusted for differences in family size and other relevant matters. Once this is done, it is only natural that people should make authoritative statements about the proportion of "poor families" having this or that characteristic, as if "poor families" could be unmistakably recognized by some birthmark on the forehead. The real mischief begins, however, when we assume that we have corralled and contained the poverty problem in a statistical isolation ward. The next step is to proclaim that the percentage of families in poverty is 42 percent lower than it was in 1950. Then we predict that our vigorous assault on poverty will eliminate the evil altogether by 1975. By this time we have forgotten that we are talking about family incomes of three thousand 1958 dollars rather than poverty in any real sense. Is it really plausible that poverty, as a social condition, would have the same meaning in 1975 as in 1950? If this were true, then presumably poverty would have the same meaning in the United States as in Europe; but we know very well that in some European countries where poverty is said to have been virtually abolished, most of the families have equivalent to less than $3000 in 1958 dollars. Furthermore, income level is not the only dimension of true poverty even at one time and place.

If poverty were strictly a matter of physical deprivation, a lack of means for essential subsistence, a static definition would serve the purpose. But since our ancestors once lived in caves without feeling impoverished, we must recognize that poverty has an important relative or comparative aspect. A family is poor if it cannot come even close to the accepted community standards of income and consumption. I am not referring to the "good life" but only to the "decent life." A family without two cars and two bathrooms may feel sorry for itself but we need not regard it as poor. But a family with no car and no bathroom, no telephone, and no daily paper must be considered poor, even if there is enough food and clothing to get along on, because it is not sharing in the basic prerequisites of American society.

If poverty is defined in relation to community norms, it follows that there must be significant psychological and social attributes. These are recognized by the more perceptive writers on the subject. In an achievement-oriented society, a poor man is one who feels inadequate and ashamed because he has not been able to make the grade. The poor are out of things: they receive no mail, they go to no meetings, they have no social life, they never get their names in the paper except when they are arrested.

I am not suggesting that statisticians should be blamed for failing to measure the subjective, social, and spiritual aspects of poverty. But perhaps they do have some responsibility to warn the layman against the danger of confusing the shadow with the substance.

A better-known example of this type of confusion is found in the late Dr. Kinsey's research on sexual behavior. Kinsey made elaborate measurements of the one dimension of sexual behavior most easily measured. Almost immediately it was taken for granted that he was dealing in some significant way not only with this quantifiable unit of measurement but also with sexuality and sexual fulfillment. This notion that sex is quantitative rather than qualitative has led to the most serious and widespread mental and moral disorientation in the United States. Something called sex can indeed be packaged and merchandized, promoted and advertised, measured and maximized. But is it the real thing?

The hurtfulness of substituting a simple statistical measure for a complex underlying reality is aggravated when the statistical measure is based on a questionable analogy. One of the more questionable analogies in current circulation describes man as a "human resource" and education as an "investment in human resources." These locutions are harmless enough when employed merely to argue that awakening countries should educate their people in order to "develop human resources," something akin to iron ore deposits; or that retraining should be provided unemployed workers in order to "overcome obsolescence of human resources," or that pension plans should be established to make allowances for "depreciation of the human machine." So far, so good. But let us not take our analogies too seriously. Let us not suppose that any civilized country could really decide how much education is desirable by comparing the costs and benefits in financial terms at the margin of choice. Let us not fall into the error of some economists who seem to be saying what they do not really believe: that men should have jobs because this contributes to the GNP, thus improving the growth rate; and that the Great Society programs are desirable beause they add to the effective manpower supply and are therefore conducive to cost-price stability.

Even these misuses of the human-resource analogy may not seem too bad because the economists are supporting benevolent policies even in their peculiar thin-blooded fashion. But when man is viewed as an object rather than a subject, as a means rather than an end, there danger lurks. In the contemporary United States the dangers are still latent. Men still express themselves and present their demands as men. Yet we can see elsewhere in the world the barbarities that are perpetrated when the human-resource concept is carried to its logical conclusion. We can read of them in our own history and that of Western Europe. Let us keep these episodes in mind so that dangerous analogies can be held in check. Let us make sure that statistics remain the servant of immeasurable, unquantifiable man and not become the master and enemy.

I have been emphasizing the subjective or creative character of the concepts that we select to represent social processes and relationships. I have also noted the danger of identifying these simple-minded dimensions with elusive and protean reality. My final thought is that we statisticians must recognize that society is unwilling to have us measure many things that are potentially measurable. I recall an article by a former staff member of *Time Magazine*. He stated that every *Time* story is basically an editorial surrounded by factual and statistical camouflage. As the reporter writes his story, he leaves many blanks to be filled in by the Editorial Researchers. (These are listed in the eleventh tier from the top of *Time's* masthead, preceded by the chairman, the Publisher, the Editor-in-Chief, the Managing Editor, the Editors, and all the other angels and archangels, powers, and principalities.) Now this writer wrote a story beginning as follows: "There are . . . trees in Russia." The editors were angry and disappointed when the editorial researchers were unable to fill in the blank. There was a spell of heavy weather. Finally some creative individual supplied a rough estimate of the number of trees in Russia. Pleased with this evidence of positive thinking, the editors smiled and the crisis evaporated.

I am told that someone wrote a "labor speech" for President Roosevelt during the 1940 campaign and included the following statement "There are . . . collective bargaining agreements in the United States." Secretary Perkins requested the Bureau of Labor Statistics to oblige. The Commissioner asked the Deputy Commissioner, who asked the appropriate Associate Commissioner, who asked the appropriate Assistant Commissioner, who asked the appropriate Division Chief, who asked the appropriate Branch Chief, who asked the appropriate Section Chief. Now the problem was on the floor of the workshop and there was nowhere to run. Finally someone thought that 50,000 might be as good an estimate as any other. This intelligence was forwarded to the President through the Branch Chief, the Division Chief, the Assistant Commissioner, the Associate Commissioner, the Commissioner, and the Secretary; and with such authoritative support, how could anyone doubt its authenticity? Fifty thousand it was, and 50,000 it remained in countless textbooks, monographs, and magazine articles.

If we can measure the number of trees in Russia and the number of labor contracts in the United States, then why not measure changes in the quality of the hundreds of goods and services that are priced in computing the Consumer Price Index and the Wholesale Price Index? This is currently the subject of some controversy. During the first week in December, a distinguished

Governor of the Federal Reserve System testified to the Joint Economic Committee that the 1.8 percent year-to-year increase in consumer prices reported by the Bureau of Labor Statistics was only a mirage. "Concealment of quality changes has meant that prices have appeared to rise, but they haven't risen if you take quality into account," he said. During the same week a well-known financial publication learned to its surprise that the Bureau was discounting the higher prices of new 1966 automobiles for the production cost of several safety features that had become standard equipment. This was condemned as the rankest kind of statistical manipulation, showing that "prices are what you make them" in the BLS.

The quality problem provides an apt illustration of the limits of measurability, those inherent in the case as well as those enforced by circumscribed resources. The Bureau does make allowance for quality improvements in automobiles and other durable consumer goods when dollar prices are rising. The durable consumer goods are easiest because at least some quality improvements are easily identified: seat belts, frostless freezer compartments, color television. This is concededly a most limited program and I am confident that we can do better. Probably some approximate allowances could be made for better serviceability of fabrics, more expeditious release of hospital patients, and so on. But imagine what a vast enterprise would be necessary to measure *all* significant changes in the quality of consumer goods and services. Is the greater pain-killing efficacy of an improved analgesic equal to the price increase, or higher or lower, and to what extent? If appliance repairs are more unsatisfactory and more delayed than usual, how can this be quantified? What about the "feel" of a more expensive shirt and other aesthetic satisfactions? Suppose the quality improvements in a new car involve expensive repair and maintenance costs? Suppose that the outmoded, cheaper version of a product is no longer available, but some customers would prefer to have it?

This is not to argue that quality changes should be ignored. It is not to deny that a better job can be done in measuring them. I expect that more of them can be taken into account, and that improvements can be made in the present method of "linking in" new products at par. But unless Congress is willing to endow a vast consumer-research undertaking, an accurate accounting of all quality variations is like an accurate census of trees in Russia. There is no reason why it can't be done, but society may feel that it's not worth the trouble.

Some economists have argued for "hedonic" price indexes which reflect subjective consumer satisfaction. This is obviously a different kettle of fish from objective appraisal of quality changes. I do not object to the proposition that the Consumer Price Index should be a "welfare index" showing "the cost of maintaining a constant level of utility" if this proposition is advanced as an argument against an excessively rigid "fixed market basket" of goods and services over a ten-year span. Some reasonable inferences from changes in consumer behavior can be helpful in developing a realistic measure of how consumers are affected by prices. But let us beware against pretentious expectations, lest we tempt the gods. Economic theorists have never been able to deal with the concept of utility except on the plane of nominalism. From objective consumer behavior they make inferences about subjective states of mind. These inferences are generally based on the shakiest and most shallow psychological assumptions. Although the economists may feel that they are "measuring utility," they are only saying that if the consumer is willing to pay a high price for some article, he must want it pretty badly. While this may be an impressive insight, it hardly constitutes a real breakthrough into the world of objective consumer satisfaction.

If we seriously undertook to measure the total impact of the automobile on the quality of life in a subjective sense, we would certainly go far beyond the differentials in weight, length, and horsepower which have been correlated with price differentials in some competent statistical studies. "Hedonics" is defined as the branch of ethics dealing with pleasure. It follows that a true hedonic price index for automobiles would be adjusted not only for weight, length, and horsepower but also for secondary consequences thereof including smog, accident casualties, and traffic jams.

I have been urging that professional statisticians should not overestimate the extent to which they can grasp and penetrate the underlying phenomena that they seek to measure. In this respect we are no different from the other professions. Life remains an elusive mystery. The nature of true health is a mystery despite the great advances in medicine. Law and order, peace, and justice are still mysteries despite the best efforts of the legal profession. I expect that more exacting tasks will be assigned to the economic statistician because society is escalating its demands on the economy. If more stringent tests of economic performance are to be met, obviously we must develop more knowledge and use it more effectively. I expect that we will be in a better position to perform these tasks because of constant improvements in theory and technology, particularly those associated with the electronic computer. I have no doubt that society will give us more resources to work with. Yet when all is said and done, I think we will still be only scratching the surface of human life and society. At least I hope so.

Sources and Errors of Economic Statistics

OSKAR MORGENSTERN

Oskar Morgenstern is Professor of Economics at Princeton University. This article is taken from his book, *On The Accuracy of Economic Observations*.

1. LACK OF DESIGNED EXPERIMENTS

Economic statistics are not, as a rule, the result of designed experiments, although one of the earlier great economists, J. H. von Thünen, conducted careful experiments in administering his estate, kept extensive records of his operations which he then analyzed, thereby anticipating much of the later marginal utility theory. But in general, economic statistics are merely by-products or results of business and government activities and have to be taken as these determine. Therefore, they often measure, describe, or simply record something that is not exactly the phenomenon in which the economist would be interested. They are often dependent on legal rather than economic definitions of processes.

A significant difference between the use of data in the natural and social sciences is that in the former the *producer* of the observations is usually also their *user*. If he does not exploit them fully himself, they are passed on to others who, in the tradition of the sciences, are precisely informed about the origin and the manner of obtaining these data. Furthermore, new data have to be fitted into a vast body of data that have been tested over and over again and into theories that have passed through the crucible of application. Also, the quality of the work of the observers is well known, and this contributes to establishing a level of precision of and confidence in the information. In the natural sciences, even the most abstract theorists are exceedingly well informed about the precise nature, circumstances, and limitations of experiments and measurements. Indeed, without such knowledge their work would be entirely impossible or meaningless.

In the social sciences the situation is quite different. It is not often feasible to be aware of the detailed nature of the data. Summarization of data is often performed by widely separated statistical workers who are likewise far removed from the later users. And finally, the tradition has simply not yet fully established itself for the users to insist upon being fully informed about all steps of the gathering and computing of statistics. Anyone who has used economic statistics, even when prepared by the finest economic-statistical institutions, knows how exceedingly difficult it is to reestablish the conditions under which they were collected, their domain, the precise activity they define, etc. although it may be decisive to be fully informed about these various stages. One of the main reasons for this difficulty is that economic data as a rule have to cover long periods of time in order to be useful. It is rarely the case that single pieces of information, not concerned with processes that extend into the past and are likely to continue into a indefinite future, are of value for economic analysis. Thus economic data are normally time series, i.e., numbers of the same kind of event, say the price of bread, strung out over time. When the series are long, as they ought to be, it is often exceedingly difficult to know how the data were obtained in the past and to what extent temporal comparability is assured.

Many producers of primary statistics make a considerable effort to inform the reader of the details of composition, stages of classification, and all other characteristics of the statistics. There are too many cases, however, where this description is sketchy and where large gaps remain. Sometimes this is due to negligence and the belief that the authority of the reporting agency is great enough to inspire full confidence in the statistics. Such authority never exists for scientific purposes. On the other hand, the great detail involved in the collection of most economic information makes it virtually impossible physically to reproduce the entire background of the descriptive detail each time that some figures are given or used. Sometimes the official commentary to statistical tables is exceedingly lengthy and fills volumes, impossible to absorb in a manner that would lead to a correction of the given numbers by the user. By swamping the user with hosts of footnotes and explanations, the makers of statistics try to absolve themselves from the need to indicate numerical error estimates. Thus a dilemma exists that could only be overcome by the development and indication of a quantitative measure expressing the error. As will be seen, such numerical expressions are lacking at the present time; in some cases they may never become available.

The deficiency of information on procedures of data-gathering is usually less striking when *sampling* methods are used to obtain economic statistics. Although a sample may sometimes be bad and though there may be other objectionable features, their construction is subject to scientific scrutiny, and the problems that must be solved in setting up a good sample are very well known. The solutions are a function of the state of sampling theory and its application in the given case. Sampling statistics in economics—a technique that we do not discuss here any further—are fortunately gaining in importance. They suggest themselves in particular when great aggregates have to be measured, such as the determination of the volume of industrial output, share of market, sales, foreign trade, and so on. Sampling is also highly valuable in constructing price statistics. In general, it can be said that the possibilities of sampling procedures have not yet been fully utilized in economics. Wherever estimates are necessary, and often they are the only possible way to arrive at some aggregates, sampling is indispensable. This is true, for example, in estimating items in the balance of payments, such as travelers' expenditures abroad, etc., where a direct approach to totals is clearly out of the question. In addition, sampling statistics can be used as checks on complete counts in order to improve the latter. Unfortunately, not enough use is made of this opportunity.

Sampling, however, is a possible additional source of error when mistakes are made in the application of the technique. Such mistakes are sometimes exceedingly difficult to avoid and some striking instances are known, as revealed by special investigations. Furthermore, sampling statistics are susceptible to the other kinds of error derived from faulty classification, time discrepancies, poor recording, etc. Sampling errors, of course, can be estimated and are usually stated. Although they do not account for the entire error or provide a way to its numerical evaluation, the indication of this component is extremely valuable.

Even though sampling procedures are being more widely introduced, the largest masses of economic statistics simply accrue without any overall scientific design or plan. It would probably be impossible to make general plans for collecting statistics without violating some basic principles of the free exchange economy. Thus the development of economics is dependent to a very high degree upon an agglomeration of statistics that in the main is rather accidental from the point of view of economic theory.

The interplay between theory, measurement, and data collection should be as intimate in economics as it is in physics, but we are far from having reached this condition. However, the signs are multiplying that economics is moving in this direction.

2.   HIDING OF INFORMATION, LIES

There is overly often a deliberate attempt to hide information. In other words, economic and social statistics are frequently based on evasive answers and *deliberate lies* of various types. These lies arise, principally, from misunderstandings, from fear of tax authorities, from uncertainty about or dislike of government interference and plans, or from the desire to mislead competitors. Nothing of this sort occurs in nature. Nature may hold back information, is always difficult to understand, but it is believed that she does not lie deliberately. Einstein has aptly expressed this fact by saying: "Raffiniert ist der Herr Gott, aber boshaft ist er nicht." [1] In that, he follows Des-

[1] Inscription on the mantle of a fireplace in Fine Hall in Princeton University: "The Lord God is sophisticated, but not malicious."

cartes and Bacon and adheres to the classical idea of the *"vera-citas dei."* The difference between describing a statistical universe made up of physical events exclusively and one in which social events occur can be, and usually is, profound. We observe here a significant variation in the structure of the phys-ical and social sciences, provided it is true that nature is merely indifferent and not hostile to man's efforts to finding out truth—it certainly not being friendly. We shall assume indifference, though proof is, I believe, lacking.

The fact, all too frequently occurring, that statistics are slop-pily gathered and prepared at the source, for example, by the firms giving out the requested information, is a different matter altogether; it is less serious than the fact of evasion, which may or may not be present at the same time. It will be seen that the lie can also take the form of handing out literally "correct," but functionally and operationally meaningless or false statistics.

Deliberately untrue statistics offer a most serious problem with broad ramifications in the realm of statistical theory, where, however, the nature and consequences of such statistics do not seem to have been explored sufficiently. It is frequently to the advantage of business to *hide* at least some information. This is easily seen—if not directly evident—from the point of view of the theory of games of strategy. Indeed, the theory of games finds a very strong corroboration in the indisputable fact that there *are* carefully guarded business secrets. Law cannot always force correct information into the open; on the contrary, it often makes some information even more worth hiding (e.g., when taxes are imposed). The incentive to lie, or at least to hide, is also strongly influenced by the competitive situation: the more preva-lent are monopolies, quasi-monopolies, or oligopolies, the less trustworthy are many statistics deriving from those industries, especially information about prices because of secret rebates granted to different customers. Consequently, statistics derived from this kind of basic data suffer greatly in reliability. For example, where national income or personal income distribution is computed on the basis of income tax returns, the results will be of widely different accuracy for different countries, tax rates, tax morale, price movements, etc. It is well known that income tax returns for France and Italy, and probably many other

countries, have only a vague resemblance to the actual, underlying income patterns of those countries. Yet it is on the basis
of tax returns that important and elusive problems, such as the
validity of the "Pareto distribution" explaining the inequality of
personal incomes, are minutely studied. For sales—an item of
primary importance in input-output studies—it must be remembered that sales prices constitute some of the most closely
guarded secrets in many businesses. The same is often true for
inventories. A prime example is the distilling industry. There it is
vital for one company not to let any other know what its stock is,
lest it suffer in the inevitable price and market struggle.

Governments, too, are not free from falsifying statistics. This
occurs, for example, when they are bargaining with other governments and wish to obtain strategic advantages or feel im-
pelled to bluff. More often, information is simply blocked for
reasons of military security, or in order to hide the success or
failure of plans. In Fascist and Communist totalitarian countries
the suppression of statistics is often carried very far. For example,
foreign trade data are considered secret in some eastern European countries with capital punishment threatened for disclosure!
Even in the United States incomplete figures are released in
the field of atomic energy, although the known total appropria-
tions for the Atomic Energy Commission indicate that this is one
of the largest American industrial undertakings. The same
applies in this field to all present (and will apply to all future)
atomic powers. The budget for the Central Intelligence Agency,
unquestionably running into hundreds of millions of dollars, is
hidden in a multitude of other accounts in the Federal budget,
invalidating also those accounts. The Russian defense budget is
only incompletely known. An example of government falsification of statistics is Nazi Germany's stating its gold reserves far
below those actually available, as was revealed by later information. Or, more subtly, indexes of prices are computed and
published from irrelevant prices in order to hide a true price
movement. Central banks in many countries, the venerable Bank
of England not excepted, have for decades published deliberately
misleading statistics, as, for example, when part of the gold in
their possession is put under "other assets" and only part is

shown as "gold." In democratic Great Britain before World War II, the Government's "Exchange Equalization Account" suppressed for a considerable period all statistics about its gold holdings, although it became clear later that these exceeded the amount of gold shown to be held by the Bank of England at the time. This list could be greatly lengthened. If respectable governments falsify information for policy purposes, if the Bank of England lies and hides or falsifies data, then how can one expect minor operators in the financial world always to be truthful, especially when they know that the Bank of England and so many other central banks are not?

A special study of these falsified, suppressed, and misrepresented government statistics is greatly needed and should be made. The probably deliberate over- and understatements of needs and resources in the negotiations concerned with the international food situation, the Marshall Plan, etc., offer vast opportunities for such investigations—if the truth can be found out.

When the Marshall Plan was being introduced, one of the chief European figures in its administration (who shall remain nameless) told me: "We shall produce any statistic that we think will help us to get as much money out of the United States as we possibly can. Statistics which we do not have, but which we need to justify our demands, we will simply fabricate." These statistics "proving" the need for certain kinds of help will go into the historical records of the period as true descriptions of the economic conditions of those times. They may even be used in econometric work!

The true or imagined purpose of statistics often has a great influence upon the answers (especially in designed statistics). In undeveloped areas there is often an important element of "boasting," beside a general desire to give the questioner the kind of answer he would like to hear, however remote it might be from the truth.[2]

A very modern and unusually important instance of the prob-

---

[2] It is reported that in Russia in the early 1930s the central statistical authorities had worked out "lie-coefficients" with which to correct the statistical reports according to regions, industries, etc. Nothing definite is known, but quite recently Khrushchev has accused especially Russian agricultural circles of reporting grossly false statistics.

lem of obtaining data is the problem of inspection in arms control. There, sampling would have to be used in order to discover possible evasions through secret production of arms, secret atomic tests, etc. An international inspection team would encounter great difficulties not yet resolved by modern statistics. A theory of "sampling in a hostile environment" is now under development; it would be applicable to many other situations in the social world.

Where such conditions prevail, the designer may have to hide the purpose of the statistic and the nature of the statistical procedure from the subject, who, in his turn, tries to hide the truth. This is the precise setup of a nonstrictly determined two-person game where both sides have to resort to mixed or "statistical" strategies. It is an ironic circumstance that in order to get good statistics, "statistical strategies" may have to be used.

Proper techniques of questioning will have to be worked out to produce a minimum of error under these conditions. These phenomena may also be viewed as disturbances of the subjects interrogated. They are familiar to anthropologists who find that conditions in primitive societies have changed after these have previously been visited by other anthropologists. Conditions of disturbances occur also in physical experiments, where in some well-defined cases in quantum mechanics it has been shown impossible *on principle* to obtain certain types, or rather certain combinations, of information.

The undeniable existence of an unknown but undoubtedly substantial amount of deliberately falsified information presents a unique feature for the theoretical social sciences, totally absent in the natural sciences, whether historical or theoretical.

History too has to cope with this difficulty. Falsifications are notorious there and can be found everywhere. Therefore source criticism is a highly developed technique that every student of history has to learn in detail. A large literature exists in this field and many eminent historians have contributed to it. Without this tradition, the writing of history would be entirely worthless. Clearly, it is not simple to establish a "historical fact" or else there would be little need to rewrite history as often as this is done (quite apart, of course, from the ever-changing evaluation of the past).

A good illustration of the difficulty in determining the true value of historical claims is given in the classical work of Hans Delbrück,[3] who has carefully examined most military battles fought over the centuries with a view to determining the strength of the opposing forces. It is clear that the victors have always stated the defeated to have been much stronger than they were in order to make victory impressive, and the losers vice versa in order to make defeat excusable; this often creates figures that are impossible for the same occasion. Delbrück has found, for example, that if the Greek claims regarding the strength of the Persians at Thermopylae were true, there would not even have been room for the Persian troops to occupy the battlefield. Or, given the roads of the time, the last Persian troops would have just crossed the Bosporus when the first already had arrived in Greece. In this manner it goes throughout history, even up to most recent times; and what really happened is very difficult to find out.

Other instances from fields of social statistics or fact-finding are suicide statistics. They are notoriously bad because lay coroners so frequently disagree with medical men, and because great efforts have always been made to keep the fact of suicide secret.[4] This applies also to medical statistics; for generations it was considered improper to die of cancer, hence little mention of this disease. This shows up in the very limited value of statistics of death (the records of insurance companies notwithstanding). Time series, in particular, suffer from the fact that many diseases in former years were entirely unknown to medical science, although people died of them. Thus, the "growth" of certain diseases is perhaps simply their better identification. This is notorious for mental illness. For example, there are many more mental cases in Sweden per 100,000 of population than in Yugoslavia. But this is simply due to the fact that in the former country the patient is taken care of in a hospital, whereas

[3] *Geschichte der Kriegskunst in Rahmen der Politischen Geschichte* (Berlin, 1900). A brief extract is found in his (now very rare) *Numbers in History, How the Greeks Defeated the Persians* (London, 1913).
[4] Accidents are another case where great doubts often prevail as to cause and effect. Probably most murders go undetected. For example, a very large proportion of hunting accidents are apparently murders; an investigation showing this was suppressed, however.

in the latter he vegetates as the village idiot and is not recorded as a mental case. Death certificates are very difficult to make out when death is due, as is often the case, to several causes, e.g., pneumonia following upon some other affliction. Only a few countries demand autopsies in every case; and even then death cannot always be uniquely attributed to a single cause.

The difficulties of finding out what "facts" are can clearly be seen in legal procedure. Evidence is placed before juries but the outcome of their fact-finding is notoriously uncertain. In general, the experience is that the chances of establishing a fact as such before a court of law are very small and that a prediction of the outcome of a law suit is hazardous. Many witnesses lie, sometimes perjury is discovered. Even when witnesses are truthful, or trying to be truthful, their statements are subject to all the doubts and limitations that have been brought out in a vast literature on the psychology of witnesses and the reliability of memory. It would lead too far afield to deal further with these matters here, though they do illuminate some of the difficulties of fact-finding encountered also in economics.

Without knowing the extent of the falsifications that actually occur in economic statistics, it is impossible to estimate their influence upon economic theory. But the peculiar feature remains that, if economic theory is based on observations of facts (as it ought to be), these are not only subject to ordinary errors, but in addition to the influences of deliberate falsifications. If for no other reasons, this is a severe restriction on the operational value of economics as long as the magnitude of this factor has not been fully investigated. Here is a field where thorough studies are required; they will be difficult to make, but they promise important results. The theory of games takes full cognizance of the phenomenon wherever it becomes relevant.

Falsification is difficult when it is attempted in a system or organization that is well described and understood. It is virtually impossible in a small mechanical system, though for large systems there are already doubts as to its working beyond a certain degree of reliability. To introduce a false circuit in an electronic computer would be foolish, since it is bound to be discovered. But social organizations are not nearly as well described as physical systems. Hence their working behavior cannot be predicted as precisely. This means that there are degrees of freedom of behavior that are compatible with alternative, equally plausible descriptions of the system. They need not differ profoundly. In addition, it should be noted, it is possible to prove that *there can be no complete formalization of society.* Consequently, a lie or falsification relating to some part or component of the system is exceedingly hard to discover, except by chance. Yet the chance factor itself is a necessary, constituent element of every social system. Without it "bluffing," a perfectly sound move in strategic behavior by elements (persons or firms) of a social organization, would be impossible. But it is a daily occurrence. Bluffing is an essential feature of rational strategies.

So we see that a lie or falsification has to be related to the degree of our knowledge of the framework within which it is attempted.

To give an illustration: Our knowledge of the population of a country, as established by a series of population counts, to which is added our knowledge of human reproductive ability makes it difficult to introduce in the next census willful distortions that would go beyond a certain measure. Lies—or other, ordinary errors—can be discovered, though this may be laborious. An economy is much less understood, and when a government, for example, reports to be in the possession of x millions of gold instead of the true y millions, this is very hard, if not impossible, to contradict, since both x and y may be compatible with our understanding of the economy and its workings. Experiments with individual firms have shown that many falsifications of production records cannot be discovered, even by means of the most minute money accounting controls. When it comes to the recording of prices, movement of goods (especially in international trade, inventories, etc.), the possibilities are substantially widened. Even in production the wide substitutability of one material for another makes great variations plausible. Of course, nobody would believe that a large country could have doubled its steel capacity within one year—but we do not consider such crude matters.

When an economy is in the throes of a great development, coupled with a rapidly changing technology, the scope for misrepresentation is correspondingly widened. Our knowledge of dynamic processes is necessarily inferior to our ability to describe stationary conditions. Yet the economies we deal with are now and have been for decades in a period of active, dynamic development.

In summarizing, we see that there are three principal sources of false representation: First, the observer, by making a selection as to what and how much to observe, introduces a bias that it is impossible to avoid, because a complex phenomenon can never be exhaustively described. This bias, common to all science, is of no concern here. Second, the observer may deliberately hide information or falsify his findings to suit his hypotheses or his political purposes. This occurs in historical writings, even in physical science in exceptional cases of fraud, and more frequently when economic and social statistics are used or abused in the hands of unscrupulous persons or institutions. Reference to some cases has been made above. Third, the observed distinction between social and physical observations; in the latter, this factor is absent no matter how difficult it may be to discover the facts. To account for this additional character of observations in the social field, new ideas concerning the foundations of statistics are necessary, as has been indicated above. The distinction applies to both measurable and (for the time being) nonmeasurable information or observations.

3.   THE TRAINING OF OBSERVERS

Economic statistics, even when planned in detail, are frequently not gathered by highly trained observers but by personnel collected ad hoc. This is a source of the most serious kind of mass errors. Even trained census-takers and many others engaged in field work are not "observers" in a strict scientific sense. A scientific observer is the astronomer at his telescope, the physicist recording the scatter of mesons, the biologist determining the hereditary behavior of some cells, etc.; all are themselves scientists; they do not operate through agents many times removed. Except where experiments are involved, the social sciences will never get into an equivalent position as far as the basic raw material of observations is concerned. Because of the masses of data needed this would be physically impossible.

We cannot place technically trained economists or statisticians at the gates of factories in order to determine what has been produced and how much is being shipped to whom at what prices. We will have to rely on business records, kept by men and, increasingly, by machines, none of them part of the ideally needed scientific setup as such. If properly engineered (and costs of processing are a minor consideration), these data can be useful. In the future we will be able to rely more on automatic recording devices and computers, thereby improving, but also modifying, the picture.

It is well known from sampling experience (where, if properly done, one deals with strictly, though not always *well* designed statistics!) that the response is very different depending upon the type of observer, even if the latter is trained.[5] and should be—miraculously—free of bias. Detailed knowledge of how much improvement in statistics could be obtained by training, or more training (at greater expense) is difficult to come by. Hence, the phenomenon, well known from experimental physics and astronomy, of the "personal equation" assumes very much larger proportions with less definite controls and perhaps even fundamentally different characteristics.

It could perhaps be argued that it should be possible to explore the nature of lies and the influence of training and bias of the observer thoroughly in controlled experiments. In other words, a sample would be designed which would be studied to the utmost degree; from the information thus gained one could then arrive at an evaluation of these factors, even in cases where no thorough exploration would be possible. It is to be doubted, however, that such a program can be carried out at the present state of affairs; it may even encounter systematic difficulties of a nature too deep to be discussed here.

[5] For example, population statistics often show concentrations at the rounded-off ages of 20, 25, 30, etc., which are in clear contradiction to earlier information. In other words, people prefer to indicate these ages, rather than their true ones which lie in between. The response to questions about attendance at college depends to a high degree upon the social status of the questioned and that of the investigator. If he appears to belong to the college-educated class, he will more often than not hear that the questioned went to college too, and vice versa. Some of these answers are also motivated by the fact that the questioned wishes to please the interrogator.

4.   ERRORS FROM QUESTIONNAIRES

Designed economic and social statistics often require the use of questionnaires. Some are presented orally; others require written answers. Some of the latter may at times contain several hundred questions directed to the same firm or individual. Errors can and do derive from the setting up of the questionnaires and from the answers. The questions should always be so formulated that unique answers are possible. But this is often not the case, for, on the contrary, many questions are not stated unambiguously or they require more intelligence for correct answers than is possessed by the person questioned. When large numbers of questions are involved, the possibility that contradictions will occur in the answers may be great, while at the same time significant omissions may be made. Often words are used that have emotional or political connotations and prejudice the answer, depending on the individual to whom they are presented. Some questions invite evasion, lies, and, when very numerous, a summary response (sometimes capricious) in order to save time, money, and generally to avoid bother and trouble. It also makes a great deal of difference whether the same questions are presented orally, in writing, by mail, and so forth.

As is well known, each form of interrogation produces its own kind of bias. The process of asking questions and getting answers is a delicate psychological one. Apart from lying and refusal to give information, there is forgetfulness, prompting by the questioner with its own consequent bias, lack of comprehension of the question, etc. These phenomena have been studied in the literature.[6] Certain investigations of business decisions by means of questionnaires have produced results contradictory to

[6] Compare F. F. Stephan, "Sampling Opinions, Attitudes and Wants," *Proceedings of the American Philosophical Society*, Vol. 92 (February 1948), pp. 387–98; and F. F. Stephan and P. J. McCarthy, *Sampling Opinions, An Analysis of Survey Procedures* (New York 1958), which gives a comprehensive and up-to-date discussion of the difficulties and the current ways to overcome them.

A particularly interesting work is S. L. Payne, *The Art of Asking Questions* (Princeton, 1951; Rev. Ed., 1955) which shows the many ambiguities in questions asked, the different associations they evoke, and what dangerous manipulations are possible.

expectations. In these cases, however, it is difficult to arrive at a conclusion, largely because it need not be assumed that business men always are able to interpret their own actions. A human being, though a living organism, is not necessarily able to describe his own functioning; yet he has a formidable "experience" of living. It takes several sciences to describe the process of living and to tell how the human body functions, and these sciences clearly have not yet come to the end of their questions and the search for answers.

The field of questionnaires is comparatively new in statistics, and the theory covering it is far from completely developed. Indeed, it is doubtful that even the qualitative description and enumeration of its characteristics are complete. Here we merely point to its existence and emphasize its enormous importance, especially for those large data collections connected with input-output studies of industry and business, determination of incomes, spending habits, etc.

The difficulties of preparing good questionnaires and using them properly are, indeed, formidable but not appreciated by the public. The simple fact is that it is not easy to ask good questions and to insure that intelligent, reliable, and honest answers will be given. Science is, after all, nothing but a continuing effort to find the right questions, followed by the search for answers. And the question is often more important than the answer. It is not different in drawing up questionnaires for economic matters. Progress in science has often been blocked by having asked the wrong question. When a field such as economics depends so largely on asking human rather than inanimate nature, the problem of the right question assumes new importance.

There is one requirement that can and should be fulfilled, whatever the state of the theory that covers the problems arising from the design and use of questionnaires: whenever questionnaires are used (or questions are asked orally), their precise text, with instructions for use, should be published together with the final results and interpretation. A mere paraphrasing of the question is insufficient because it may involve subtle changes in the meaning and undertones. In that respect, however, many producers of primary statistics, including government agencies, fail to comply and give only the vaguest kind of information about the underlying questions. This circumstance deprives the user of the results of a great deal of their value. Publicly used statistics for which the user does not have access to this information should be rejected, no matter how interesting and important the particular field may be. On the other hand, the publication of the frequently very numerous and complicated questions does not make the use of the answers any easier, because the reader is supposed to accomplish a difficult task of interpretation and evaluation for which he may not always be prepared.

Countless examples could be given of poorly designed questionnaires and samples. But we are not looking for the inadequate in economic statistics. Rather we try to assess the presumably best and to find out how much confidence can be placed in the work of the most renowned institutions. When troublesome errors are found there we have to conclude that elsewhere they will not be much different.

An interesting example, pertaining to questionnaires but pointing toward wider problems, is the following: In 1953, the British Ministry of Labor and National Service conducted an inquiry into household expenditure by questionnaire and by interview. A total of 20,000 households were asked to list all their expenditures over a period of three weeks. Of the returns, only 12,911 were usable. The figures were broken down in many ways, one way being expenditure on various items by household against income of head of household. We can extract from Table 9 of the "Report of an Enquiry into Household Expenditure in 1953–54," (H.M.S.O., London), the figures indicated in our Table 1. We note the huge figure for weekly expenditure on women's outer clothing for the richest group (over £50). However, a footnote gives the reason for this: "One member of a household in this group spent £1903 on one item during the period"—the item presumably being a very expensive fur coat. This fur coat keeps reappearing throughout other tables and each time provides us with a ridiculous figure. There is nothing *wrong* with the data; and the statisticians who wrote the report have been perfectly honest, but their results would have been more useful if the coat had been left out.

This example shows, incidentally, with what great care cer-

TABLE 1

WEEKLY INCOME OF HEAD OF HOUSEHOLD

	£50 or more	£30 to £50	£20 to £30	£14 to £20	£10 to £14	£8 to £10	£6 to £8	£3 to £6	Under £3
Number in sample falling in this group	58	111	291	1003	2765	2779	2472	1589	1843
	s. d.	s. d.	s. d.	s. d.	s. d.	s. d.	s. d.	s. d.	s. d.
Weekly expenditure on women's outer clothing	225 6.3	11 4.9	13 9.3	9 8.2	6 2.0	4 9.6	4 11.7	4 8.9	3 5.7

(s = shillings, d = pence)

tain statistical phenomena have to be treated. When they are encountered and recognized, they give rise to refinements in statistical methodology that constitute a progress of our understanding of such situations. They also make it clear that there always have existed errors of a more elusive kind which now can be avoided. But the approach to these problems also puts ever-increasing demands on the user of statistics that cannot always be met.

Specifically, the above example shows that it is dangerous to break down results of surveys into many, very small groups. The number of "subjects" lying in some of these will be small, and the results will be inaccurate. These "outlyers" are rare events (in terms of the sample) and probably belong to a different statistical distribution than the one encountered. They can mislead badly and will do so unless there is an immediate, intuitive reason to recognize the circumstances, as in the case of the fur coat. Even if its value had only been one-tenth it would still have biased the statistics, but this would not have been as obvious.

Techniques for the rejection of outlyers have been developed. They show that outlyers appear in many statistics and can lead to important inaccuracies, unless good methods for their rejection are used. Poor methods can produce other biases, hard to discover.

We have introduced this matter in order to show that one may sometimes spot an obvious or striking fact and recognize it as an error or distortion; but behind it there usually are many more of the same kind, yet hidden and elusive. They can be brought to light only by sophisticated statistical theory.

How to Lie
With Statistics

DARRELL HUFF

Darrell Huff is a partner in Cavedale Craftsmen. This
article comes from his well-known book of the same name,
illustrated by Irving Geis and published in 1954.

1. THE WELL-CHOSEN AVERAGE

You, I trust, are not a snob, and I certainly am not in the
real-estate business. But let's say that you are and I am and that
you are looking for property to buy along a road that is not far
from the California valley in which I live.

Having sized you up, I take pains to tell you that the average
income in this neighborhood is about $15,000 a year. Maybe that
clinches your interest in living here; anyway, you buy and that
handsome figure sticks in your mind. More than likely, since we
have agreed that for the purposes of the moment you are a bit of
a snob, you toss it in casually when telling your friends about
where you live.

A year or so later we meet again. As a member of some tax-
payers' committee, I am circulating a petition to keep the tax
rate down or assessments down or bus fare down. My plea is
that we cannot afford the increase: After all, the average income
in this neighborhood is only $3500 a year. Perhaps you go along
with me and my committee in this—you're not only a snob, you're
stingy too—but you can't help being surprised to hear about that
measly $3500. Am I lying now, or was I lying last year?

You can't pin it on me either time. That is the essential beauty of doing your lying with statistics. Both those figures are legitimate averages, legally arrived at. Both represent the same data, the same people, the same incomes. All the same it is obvious that at least one of them must be so misleading as to rival an out-and-out lie.

My trick was to use a different kind of average each time, the word "average" having a very loose meaning. It is a trick commonly used, sometimes in innocence but often in guilt, by fellows wishing to influence public opinion or sell advertising space. When you are told that something is an average you still don't know very much about it unless you can find out which of the common kinds of average it is—mean, median, or mode.

The $15,000 figure I used when I wanted a big one is a mean, the arithmetic average of the incomes of all the families in the neighborhood. You get it by adding up all the incomes and dividing by the number there are. The smaller figure is a median, and so it tells you that half the families in question have more than $3500 a year and half have less. I might also have used the mode, which is the most frequently met-with figure in a series. If in this neighborhood there are more families with incomes of $5000 a year than with any other amount, $5000 a year is the modal income.

In this case, as usually is true with income figures, an unqualified "average" is virtually meaningless. One factor that adds to the confusion is that with some kinds of information all the averages fall so close together that, for casual purposes, it may not be vital to distinguish among them.

If you read that the average height of the men of some primitive tribe is only five feet, you get a fairly good idea of the stature of these people. You don't have to ask whether that average is a mean, median, or mode; it would come out about the same. (Of course, if you are in the business of manufacturing overalls for Africans, you would want more information than can be found in any average. This has to do with ranges and deviations, and we'll tackle that one in the next section.)

The different averages come out close together when you deal with data, such as those having to do with many human characteristics, that have the grace to fall close to what is called the normal distribution. If you draw a curve to represent it you get something shaped like a bell; and mean, median, and mode fall at the same point.

Consequently one kind of average is as good as another for describing the heights of men, but for describing their pocketbooks it is not. If you should list the annual incomes of all the families in a given city you might find that they ranged from not much to perhaps $50,000 or so, and you might find a few very large ones. More than 95 percent of the incomes would be under $10,000, putting them way over toward the left-hand side of the curve. Instead of being symmetrical, like a bell, it would be skewed. Its shape would be a little like that of a child's slide, the ladder rising sharply to a peak, the working part sloping gradually down. The mean would be quite a distance from the median. You can see what this would do to the validity of any comparison made between the "average" (mean) of one year and the "average" (median) of another.

In the neighborhood where I sold you some property the two averages are particularly far apart because the distribution is markedly skewed. It happens that most of your neighbors are small farmers or wage earners employed in a nearby village or elderly retired people on pensions. But three of the inhabitants are millionaire week-enders and these three boost the total income, and therefore the arithmetic average, enormously. They boost it to a figure that practically everybody in the neighborhood has a good deal less than. You have in reality the case that sounds like a joke or a figure of speech: nearly everybody is below average.

That's why when you read an announcement by a corporation executive or a business proprietor that the average pay of the people who work in his establishment is so much, the figure may mean something and it may not. If the average is a median, you can learn something significant from it: half the employees make more than that; half make less. But if it is a mean (and believe me it may be that if its nature is unspecified) you may be getting nothing more revealing than the average of one $45,000 income—the proprietor's—and the salaries of a crew of underpaid workers. "Average annual pay of $5700" may conceal both the $2000 salaries and the owner's profits taken in the form of a whopping salary.

Let's take a longer look at that one. The facing page shows
how many people get how much. The boss might like to express
the situation as "average wage $5700," using that deceptive
mean. The mode, however, is more revealing: the most common
rate of pay in this business is $2000 a year. As usual, the median
tells more about the situation than any other single figure does;
half the people get more than $3000 and half get less.

How neatly this can be worked into a whipsaw device in
which, the worse the story, the better it looks is illustrated in
some company statements. Let's try our hand at one in a small
way.

You are one of the three partners who own a small manufacturing business. It is now the end of a very good year. You have
paid out $198,000 to the ninety employees who do the work of
making and shipping the chairs or whatever it is that you manu-
facture. You and your partners have paid yourselves $11,000
each in salaries. You find there are profits for the year of $45,000
to be divided equally among you. How are you going to describe
this? To make it easy to understand, you put it in the form of
averages. Since all the employees are doing about the same kind
of work for similar pay, it won't make much difference whether
you use a mean or a median. This is what you come out with:

Average wage of employees	$ 2,200
Average salary and profit of owners	26,000

That looks terrible, doesn't it? Let's try it another way. Take
$30,000 of the profits and distribute it among the three partners
as bonuses. And this time, when you average up the wages, include yourself and your partners. And be sure to use a mean.

Average wage or salary	$2806.45
Average profit of owners	5000.00

Ah. That looks better. Not as good as you could make it look,
but good enough. Less than 6 percent of the money available for
wages and profits has gone into profits, and you can go further
and show that, too, if you like. Anyway, you've got figures now
that you can publish, post on a bulletin board, or use in bargaining.

This is pretty crude because the example is simplified, but it

THE WELL-CHOSEN AVERAGE

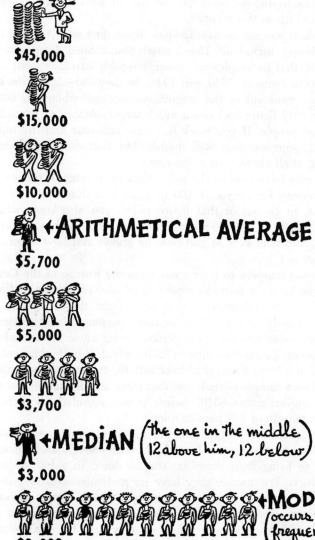

$45,000

$15,000

$10,000

←ARITHMETICAL AVERAGE
$5,700

$5,000

$3,700

←MEDIAN (the one in the middle)
              12 above him, 12 below)
$3,000

←MODE (occurs most frequently)
$2,000

is nothing to what has been done in the name of accounting. Given a complex corporation with hierarchies of employees ranging all the way from beginning typist to president with a several-hundred-thousand-dollar bonus, all sorts of things can be covered up in this manner.

So when you see an average-pay figure, first ask: Average of what? Who's included? The United States Steel Corporation once said that its employees' average weekly earnings went up 107 percent between 1940 and 1948. So they did—but some of the punch goes out of the magnificent increase when you note that the 1940 figure includes a much larger number of partially employed people. If you work half-time one year and full-time the next, your earnings will double, but that doesn't indicate anything at all about your wage rate.

You may have read in the paper that the income of the average American family was $3100 in 1949. You should not try to make too much out of that figure unless you also know what "family" has been used to mean, as well as what kind of average this is. (And who says so and how he knows and how accurate the figure is.)

This one happens to have come from the Bureau of the Census. If you have the Bureau's report you'll have no trouble finding the rest of the information you need right there: This is a median; "family" signifies "two or more persons related to each other and living together." (If persons living alone are included in the group, the median slips to $2700, which is quite different.) You will also learn if you read back into the tables that the figure is based on a sample of such size that there are 19 chances out of 20 that the estimate—$3107 before it was rounded—is correct within a margin of $59 plus or minus.

That probability and that margin add up to a pretty good estimate. The census people have both skill enough and money enough to bring their sampling studies down to a fair degree of precision. Presumably they have no particular axes to grind. Not all the figures you see are born under such happy circumstances, nor are all of them accompanied by any information at all to show how precise or unprecise they may be. We'll work that one over in the next section.

Meanwhile you may want to try your skepticism on some

items from "A Letter from the Publisher" in *Time* magazine. Of new subscribers it said, "Their median age is 34 years and their average family income is $7270 a year." An earlier survey of "old TIMErs" had found that their "median age was 41 years. . . . Average income was $9535. . . ." The natural question is why, when median is given for ages both times, the kind of average for incomes is carefully unspecified. Could it be that the mean was used instead because it is bigger, thus seeming to dangle a richer readership before advertisers?

2.   THE LITTLE FIGURES THAT ARE NOT THERE

Users report 23 percent fewer cavities with Doakes' toothpaste, the big type says. You could do with 23 percent fewer aches, so you read on. These results, you find, come from a reassuringly "independent" laboratory, and the account is certified by a certified public accountant. What more do you want?

Yet if you are not outstandingly gullible or optimistic, you will recall from experience that one toothpaste is seldom much better than any other. Then how can the Doakes people report such results? Can they get away with telling lies, and in such big type at that? No, and they don't have to. There are easier ways and more effective ones.

The principal joker in this one is the inadequate sample, statistically inadequate, that is; for Doakes' purpose it is just right. That test group of users, you discover by reading the small type, consisted of just a dozen persons. (You have to hand it to Doakes, at that, for giving you a sporting chance. Some advertisers would omit this information and leave even the statistically sophisticated only a guess as to what species of chicanery was afoot. His sample of a dozen isn't so bad either, as these things go. Something called Dr. Cornish's Tooth Powder came onto the market a few years ago with a claim to have shown "considerable success in correction of . . . dental caries." The idea was that the powder contained urea, which laboratory work was supposed to have demonstrated to be valuable for the purpose. The pointlessness of this was that the experimental work had been purely preliminary and had been done on precisely six cases.)

But let's go back to how easy it is for Doakes to get a headline without a falsehood in it and everything certified at that. Let any small group of persons keep count of cavities for six months, then switch to Doakes'. One of three things is bound to happen: distinctly more cavities, distinctly fewer, or about the same number. If the first or last of these possibilities occurs, Doakes & Company files the figures (well out of sight somewhere) and tries again. Sooner or later, by the operation of chance, a test group is going to show a big improvement worthy of a headline and perhaps a whole advertising campaign. This will happen whether they adopt Doakes' or baking soda or just keep on using their same old dentifrice.

The importance of using a small group is this: with a large group any difference produced by chance is likely to be a small one and unworthy of big type. A 2 percent improvement claim is not going to sell much toothpaste.

How results that are not indicative of anything can be produced by pure chance—given a small enough number of cases—is something you can test for yourself at small cost. Just start tossing a penny. How often will it come up heads? Half the time, of course. Everyone knows that.

Well, let's check that and see. . . . I have just tried 10 tosses and got heads 8 times, which proves that pennies come up heads 80 percent of the time. Well, by toothpaste statistics they do. Now try it yourself. You may get a fifty-fifty result, but probably you won't; your result, like mine, stands a good chance of being quite a ways away from fifty-fifty. But if your patience holds out for a thousand tosses you are almost (though not quite) certain to come out with a result very close to half heads—a result, that is, which represents the real probability. Only when there is a substantial number of trials involved is the law of averages a useful description or prediction.

How many is enough? That's a tricky one too. It depends among other things on how large and how varied a population you are studying by sampling. And sometimes the number in the sample is not what it appears to be.

A remarkable instance of this came out in connection with a test of a polio vaccine a few years ago. It appeared to be an impressively large-scale experiment as medical ones go: 450 children were vaccinated in a community and 680 were left unvaccinated, as controls. Shortly thereafter the community was visited by an epidemic. Not one of the vaccinated children contracted a recognizable case of polio.

Neither did any of the controls. What the experimenters had overlooked or not understood in setting up their project was the low incidence of paralytic polio. At the usual rate, only two cases would have been expected in a group this size, and so the test was doomed from the start to have no meaning. Something like 15 to 25 times this many children would have been needed to obtain an answer signifying anything.

Many a great, if fleeting, medical discovery has been launched similarly. "Make haste," as one physician put it, "to use a new remedy before it is too late."

The guilt does not always lie with the medical profession alone. Public pressure and hasty journalism often launch a treatment that is unproved, particularly when the demand is great and the statistical background hazy. So it was with the cold vaccines that were popular some years back, and the antihistamines more recently. A good deal of the popularity of these unsuccessful "cures" sprang from the unreliable nature of the ailment and from a defect of logic. Given time, a cold will cure itself.

How can you avoid being fooled by unconclusive results? Must every man be his own statistician and study the raw data for himself? It is not that bad; there is a test of significance that is easy to understand. It is simply a way of reporting how likely it is that a test figure represents a real result rather than something produced by chance. This is the little figure that is not there—on the assumption that you, the lay reader, wouldn't understand it. Or that, where there's an axe to grind, you would.

If the source of your information gives you also the degree of significance, you'll have a better idea of where you stand. This degree of significance is most simply expressed as a probability, as when the Bureau of the Census tells you that there are 19 chances out of 20 that their figures have a specified degree of precision. For most purposes nothing poorer than this 5-percent level of significance is good enough. For some, the demanded level is 1 percent, which means that there are 99 chances out of 100 that an apparent difference, or whatnot, is real. Anything this likely is sometimes described as "practically certain."

There's another kind of little figure that is not there, one whose absence can be just as damaging. It is the one that tells the range of things or their deviation from the average that is given. Often an average—whether mean or median, specified or unspecified—is such an oversimplification that it is worse than useless. [Knowing nothing about a subject is frequently healthier than knowing what is not so, and a little learning may be a dangerous thing.]

Altogether too much of recent American housing, for instance, has been planned to fit the statistically average family of 3.6 persons. Translated into reality this means three or four persons, which, in turn, means two bedrooms. And this size family, "average" though it is, actually makes up a minority of all families. "We build average houses for average families," say the builders—and neglect the majority that are larger or smaller. Some areas, in consequence of this, have been overbuilt with two-bedroom houses, underbuilt in respect to smaller and larger units. So here is a statistic whose misleading incompleteness has had expensive consequences. Of it, the American Public Health Association says: "When we look beyond the arithmetical average to the actual range which it misrepresents, we find that the three-person and four-person families make up only 45 percent of the total. Thirty-five percent are one-person and two-person; 20 percent have more than four persons."

Common sense has somehow failed in the face of the convincingly precise and authoritative 3.6. It has somehow outweighed what everybody knows from observation: that many families are small and quite a few are large.

In somewhat the same fashion those little figures that are missing from what are called "Gesell's norms" have produced pain in papas and mamas. Let a parent read, as many have done in such places as Sunday rotogravure sections, that "a child" learns to sit erect at the age of so many months and he thinks at once of his own child. Let his child fail to sit by the specified age and the parent must conclude that his offspring is "retarded" or "subnormal" or something equally invidious. Since half the children are bound to fail to sit by the time mentioned, a good many parents are made unhappy. Of course, speaking mathematically, this unhappiness is balanced by the joy of the other

50 percent of parents in discovering that their children are "advanced." But harm can come of the efforts of the unhappy parents to force their children to conform to the norms and thus be backward no longer.

All this does not reflect on Dr. Arnold Gesell or his methods. The fault is in the filtering-down process from the researcher through the sensational or ill-informed writer to the reader who fails to miss the figures that have disappeared in the process. A good deal of the misunderstanding can be avoided if, to the "norm" or average is added an indication of the range. Parents seeing that their youngsters fall within the normal range will quit worrying about small and meaningless differences. Hardly anybody is exactly normal in any way, just as 100 tossed pennies will rarely come up exactly 50 heads and 50 tails.

Confusing "normal" with "desirable" makes it all the worse. Dr. Gesell simply stated some observed facts; it was the parents who, in reading the books and articles, concluded that a child who walks late by a day or a month must be inferior.

A good deal of the stupid criticism of Dr. Alfred Kinsey's well-known (if hardly well-read) report came from taking normal to be equivalent to good, right, desirable. Dr. Kinsey was accused of corrupting youth by giving them ideas and particularly by calling all sorts of popular but unapproved sexual practices normal. But he simply said that he had found these activities to be usual, which is what normal means, and he did not stamp them with any seal of approval. Whether they were naughty or not did not come within what Dr. Kinsey considered to be his province. So he ran up against something that has plagued many another observer: It is dangerous to mention any subject having high emotional content without hastily saying whether you are for or agin it.

The deceptive thing about the little figure that is not there is that its absence so often goes unnoticed. That, of course, is the secret of its success. Critics of journalism as practiced today have deplored the paucity of good old-fashioned leg work and spoken harshly of "Washington's armchair correspondents," who live by uncritically rewriting government handouts. For a sample of unenterprising journalism take this item from a list of "new industrial developments" in the news magazine *Fortnight:*

"a new cold temper bath which triples the hardness of steel, from Westinghouse."

Now that sounds like quite a development . . . until you try to put your finger on what it means. And then it becomes as elusive as a ball of quicksilver. Does the new bath make just any kind of steel three times as hard as it was before treatment? Or does it produce a steel three times as hard as any previous steel? Or what does it do? It appears that the reporter has passed along some words without inquiring what they mean, and you are expected to read them just as uncritically for the happy illusion they give you of having learned something. It is all too reminiscent of an old definition of the lecture method of classroom instruction: a process by which the contents of the textbook of the instructor are transferred to the notebook of the student without passing through the head of either party.

A few minutes ago, while looking up something about Dr. Kinsey in *Time,* I came upon another of those statements that collapse under a second look. It appeared in an advertisement by a group of electric companies in 1948. "Today, electric power is available to more than three-quarters of U.S. farms. . . ." That sounds pretty good. Those power companies are really on the job. Of course, if you wanted to be ornery, you could paraphrase it into "Almost one-quarter of U.S. farms do not have electric power available today." The real gimmick, however, is in that word "available," and by using it the companies have been able to say just about anything they please. Obviously this does not mean that all those farmers actually have power, or the advertisement surely would have said so. They merely have it "available"—and that, for all I know, could mean that the power lines go past their farms or merely within 10 or 100 miles of them.

Let me quote a title from an article published in *Collier's* in 1952: "You Can Tell *Now* HOW TALL YOUR CHILD WILL GROW." With the article is conspicuously displayed a pair of charts, one for boys and one for girls, showing what percentage of his ultimate height a child reaches at each year of age. "To determine your child's height at maturity," says a caption, "check present measurement against chart."

The funny thing about this is that the article itself—if you read on—tells you what the fatal weakness in the chart is. Not all children grow in the same way. Some start slowly and then speed up; others shoot up quickly for a while, then level off slowly; for still others growth is a relatively steady process. The chart, as you might guess, is based on averages taken from a large number of measurements. For the total, or average, heights of 100 youngsters taken at random it is no doubt accurate enough, but a parent is interested in only one height at a time, a purpose for which such a chart is virtually worthless. If you wish to know how tall your child is going to be, you can probably make a better guess by taking a look at his parents and grandparents. That method isn't scientific and precise like the chart, but it is at least as accurate.

I am amused to note that, taking my height as recorded when I enrolled in high-school military training at fourteen and ended up in the rear rank of the smallest squad, I should eventually have grown to a bare five feet eight. I am five feet eleven. A three-inch error in human height comes down to a poor grade of guess.

Before me are wrappers from two boxes of Grape-Nuts Flakes. They are slightly different editions, as indicated by their testimonials: one cites Two-Gun Pete and the other says, "If you want to be like Hoppy . . . you've got to eat like Hoppy!" Both offer charts to show ("Scientists *proved* it's true!") that these flakes "start giving you energy in two minutes!" In one case the chart hidden in these forests of exclamation points has numbers up the side; in the other case the numbers have been omitted. This is just as well, since there is no hint of what the numbers mean. Both show a steeply climbing red line ("energy release") but one has it starting one minute after eating Grape-Nuts Flakes, the other two minutes later. One line climbs about twice as fast as the other, too, suggesting that even the draftsman didn't think these graphs meant anything.

Such foolishness could be found only on material meant for the eye of a juvenile or his morning-weary parent, of course. No one would insult a big businessman's intelligence with such statistical tripe . . . or would he? Let me tell you about a graph used to advertise an advertising agency (I hope this isn't getting confusing) in the rather special columns of *Fortune* magazine. The line on this graph showed the impressive upward trend of

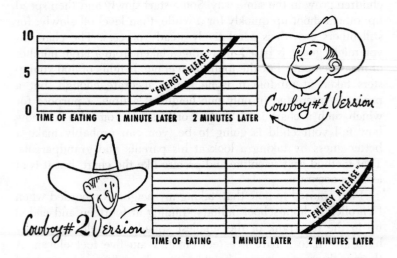

the agency's business year by year. There were no numbers. With
equal honesty this chart could have represented a tremendous
growth, with business doubling or increasing by millions of dollars a year, or the snail-like progress of a static concern adding
only a dollar or two to its annual billings. It make a striking
picture, though.

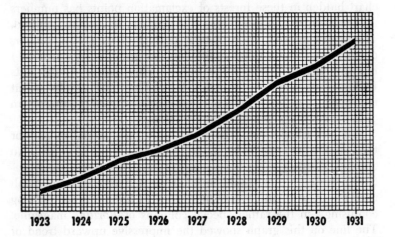

Place little faith in an average or a graph or a trend when those important figures are missing. Otherwise you are as blind as a man choosing a camp site from a report of mean temperature alone. You might take 61° as a comfortable annual mean, giving you a choice in California between such areas as the inland desert and San Nicolas Island off the south coast. But you can freeze or roast if you ignore the range. For San Nicolas it is 47 to 87° but for the desert it is 15 to 104°

Oklahoma City can claim a similar average temperature for the last 60 years: 60.2°. But as you can see from the chart below, that cool and comfortable figure conceals a range of 130°.

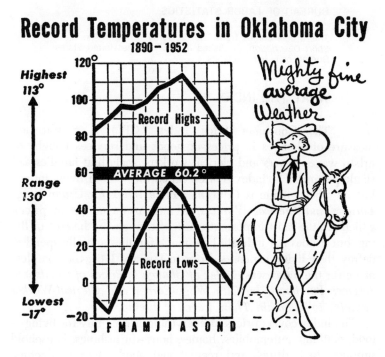

Record Temperatures in Oklahoma City
1890 – 1952

Highest 113°

Range 130°

Lowest –17°

Record Highs

AVERAGE 60.2°

Record Lows

Mighty fine average Weather

The Consumer Price Index

BUREAU OF LABOR STATISTICS

This article comes from *The Consumer Price Index: A Short Description,* published in 1967 by the United States Bureau of Labor Statistics.

WHAT THE INDEX IS

The Consumer Price Index (CPI) is a statistical measure of changes in prices of goods and services bought by urban wage earners and clerical workers, including families and single persons. The index is often called the "cost-of-living index," but its official name is *Consumer Price Index for Urban Wage Earners and Clerical Workers.* It measures changes in prices, which are the most important cause of changes in the cost of living, but it does not indicate how much families actually spend to defray their living expenses. Prior to January 1964, the complete name for the index was: *Index of Change in Prices of Goods and Services Purchased by City Wage-Earner and Clerical-Worker Families to Maintain Their Level of Living.*

The index covers prices of everything people buy for living—food, clothing, automobiles, homes, house furnishings, household supplies, fuel, drugs, and recreational goods; fees to doctors, lawyers, beauty shops; rent, repair costs, transportation fares, public utility rates, etc. It deals with prices actually charged to consumers, including sales and excise taxes. It also includes real estate taxes on owned homes, but it does not include income or personal property taxes.

Since January 1964, the index has applied to single workers living alone, as well as to families of two persons or more. The average size of families represented in the index is about 3.7 persons, and the average family income in 1960–61 was about $6250 after taxes. The average income after taxes of single persons represented in the index was about $3560.

THE MEANING OF THE INDEX MEASUREMENT

The index measures price changes from a designated reference period. Since 1962, the base reference period for the CPI has been the average of three years—1957, 1958, and 1959—as 100.0. (Index numbers are also available regularly on $1939 = 100$ and $1947–49 = 100$ bases, and they can be converted to any desired base period.) An index of 110 means there was a 10-percent increase in prices since the base period; similarly, an index of 90 means a 10-percent decrease.

Movements of the index from one date to another are usually expressed as percent changes rather than changes in index points, because index points are affected by the base period, while percent changes are not. The following example illustrates the difference between percent change and index points change:

Period	Index		
	Base A	Base B	Base C
I	112.5	168.8	225.0
II	121.5	182.3	243.0
Index points change	9.0	13.5	18.0
Percent change	$\frac{9.0}{112.5} \times 100 = 8.0$	$\frac{13.5}{168.8} \times 100 = 8.0$	$\frac{18.0}{225.0} \times 100 = 8.0$

The Bureau of Labor Statistics calculates a monthly index representing all urban places in the United States—The U.S. City Average Index—and a separate index for each of 23 Standard Metropolitan Statistical Areas. The individual city indexes measure how much prices have changed in a particular city from time to time; but they do not show whether prices or living costs are higher or lower in one city than in another. For example, consider the prices of a single item in two cities in two years:

| | Price | | Index, Year II |
	Year I	Year II	(Year I = 100)
City A	$0.30	$0.60	200
City B	.40	.70	175

The price is higher in City B in each of the two years, but the relative increase in price in City B is less and therefore the index is lower.

USES OF THE INDEX

The Consumer Price Index is used widely by the general public to guide family budgeting and to understand what is happening to family finances. It is used extensively in labor-management contracts to adjust wages. Automatic adjustments based on changes in the index are incorporated in some wage contracts and in a variety of other types of contracts, such as long-term leases. In addition, the CPI is used as a measure of changes in the purchasing power of the dollar for such diverse purposes as adjusting royalties, pensions, welfare payments, and occasionally alimony payments. It also is used widely as a reflection of inflationary or deflationary trends in the economy.

BRIEF HISTORY OF THE INDEX

The Bureau of Labor Statistics has been calculating the Consumer Price Index for nearly five decades. The weighting factors, the list of items included in the market basket, and the cities in which price data were collected for calculating the index have been updated several times during that period. Initially, they were based on a survey of expenditures by wage earners and clerical workers in 1917–19. Because people's buying habits changed substantially by the mid-1930s, a new study was made covering expenditures in the years 1934–36 which provided the basis for a comprehensively revised index introduced in 1940 with retroactive calculations back to 1935.

During World War II, when many commodities were scarce and goods were rationed, the index weights were adjusted to reflect these shortages. Again in 1950, the Bureau made interim adjustments, based on surveys of consumer expenditures in seven cities between 1947 and 1949, to reflect the most important effects of immediate postwar changes in buying patterns. This adjustment was followed by the first comprehensive postwar revision of the index, which was completed in January 1953. At that time, not only were the weighting factors, list of items, and sources of price data updated, but many improvements in pricing and calculation methods also were introduced.

The most recent comprehensive revision of the index was completed in January 1964. To determine the current pattern of expenditures for goods and services by wage earners and clerical workers, the Bureau made a Consumer Expenditure Survey (CES) covering the period 1960–61. The sample of cities in the survey included 72 urban areas which were chosen to represent all urban places in the United States, including Alaska and Hawaii. Only 56 of the 72 areas comprise the list of cities in which price quotations are obtained for the index. In this most recent survey, as in those conducted earlier, the BLS obtained a detailed record of the kind, qualities, and amounts of all goods and services bought by each consumer unit (family or single person living alone) and of the annual amount spent for each item. A total of 4912 urban wage-earner and clerical-worker families and 585 single workers provided such records.

THE MARKET BASKET

It is not feasible or necessary to obtain current price quotations on everything that consumers buy in order to calculate a valid index of changes in consumer prices. About 400 items have been selected objectively to compose the "market basket" for current pricing, beginning with the January 1964 indexes. Not all items are priced in every city. In order to make possible estimates of sampling error, two subsamples of items have been set up. These are priced in different cities and in different outlet samples. The list includes the most important goods and services and a sample of the less important ones. In combination, these represent all items purchased. The content of this market basket in terms of items, quantities, and qualities is kept essentially unchanged in the index calculation between major revisions so that any movement of the index from one month to the next is due solely to changes in prices. A comparison of the total cost of the market basket from period to period yields the measure of average price change.

PRICE DATA COLLECTION

Prices are obtained by personal visit to a representative sample of about 18,000 retail stores and service establishments where wage and clerical workers buy goods and services, including among the establishments chain stores, independent grocery stores, department and specialty stores, restaurants, professional people, and repair and service shops. Rental rates are obtained from about 40,000 tenants. Reporters are located both in the city proper and in suburbs of each urban area. Cooperation of reporters is completely voluntary and generally excellent.

To insure that the index reflects only changes in prices and not changes due to quantity or quality differences, the Bureau has prepared detailed specifications to describe the items of the market basket. Specially trained Bureau representatives examine merchandise in the stores to determine whether the goods and services for which they record prices conform to the specifications. Where the precisely specified item is not sold at a particular retail establishment, the Bureau's representative obtains a detailed technical description of the item on which prices are quoted, in order to insure that prices will be quoted on the same quality and quantity from time to time.

Prices are collected in each urban location at intervals ranging from once every month to once every three months, with a few items surveyed semiannually or annually. Because food prices change frequently, and because foods are a significant part of total spending, food pricing is conducted every month in each urban location. Prices of most other goods and services are collected every month in the five largest urban areas and every three months in all other places. Pricing of foods is done on three consecutive days each month; rents and items for which prices are obtained by mail are reported as of the 15th of the month; pricing of other items extends over the entire calendar month.

The Bureau uses mail questionnaires to obtain data on streetcar and bus fares, public utility rates, newspaper prices, and prices of certain other items that do not require a personal visit by Bureau agents. For a number of items, for example, home purchase, college tuition, used cars, magazines, and so forth, data collected by other government agencies or private organizations are used.

INDEX CALCULATION

A standard statistical formula [1] is used to calculate the Consumer Price Index from prices for the market basket items. Average price changes from the previous pricing period to the current month are expressed in percentage terms for each item, and the percent changes for the various goods and services are combined, using weighting factors based on the item's importance in family spending and that of other items which it represents. This composite importance is called the cost weight of the market basket item. There is a set of separate cost weights for each of the 56 urban locations included in the index. The following hypothetical example for pork illustrates the index procedure. (See Table A.) Identical results could be obtained for pork by multiplying prices each period by the implied physical quantities included in the market basket, as illustrated in Table B. The average change in pork prices is computed by comparing the sum of the cost weights in October with the comparable sum for September, as follows:

$$\frac{\text{October cost weight}}{\text{September cost weight}} \quad \frac{\$33.85}{\$33.00} \times 100 = 102.6$$

This means that pork prices in October were 102.6 percent of (or 2.6 percent higher than) pork prices in September.

Although the second method may appear simpler, in reality it is not. Deriving the implied quantity weights is an extra operation. Furthermore, the second formulation greatly complicates the handling of the numerous substitutions of reporters and items which occur constantly in repetitive index work. Conse-

[1] See Explanation of the Index Formula in the Appendix.

TABLE A

Sample item	September price	October price	Price relative Sept. = 100	September cost weight	October cost weight
Pork chops	$0.75	$0.77¼	103.00	$15.00	$15.45
Ham	0.80	0.82	102.50	8.00	8.20
Bacon	1.00	1.02	102.00	10.00	10.20
				$33.00	$33.85

TABLE B

Sample item	Implied quantity (pounds)	September price	September cost weight	October price	October cost weight
Pork chops	20	$0.75	$15.00	$0.77¼	$15.45
Ham	10	0.80	8.00	0.82	8.20
Bacon	10	1.00	10.00	1.02	10.20
			$33.00		$33.85

quently, the first method is the one actually used for the CPI. The second illustration, however, may assist the user to understand the meaning of the index mechanism.

After the cost weights for each of the items have been calculated, they are added to area totals for commodity groups and all items. The United States totals are obtained by combining area totals, with each area total weighted according to the proportion of the total wage-earner and clerical-worker population that it represents in the index based on 1960 Census figures. In this process, it is necessary to make estimates for cities in which price data are not collected in a given month. Finally, the United States totals for the current and previous months are compared to compute the average price change.

SEASONALLY ADJUSTED INDEXES

In January 1966, the Bureau initiated publication of seasonally adjusted national indexes for selected groups and subgroups of the CPI for which there is a significant seasonal pattern of price change. Previously, the Bureau had made available seasonal factors, permitting users who wished to do so to calculate seasonally adjusted indexes. Since there is no significant seasonal movement of the "all items" index, it is not seasonally adjusted. Moreover, no seasonally adjusted indexes are computed for any of the metropolitan areas. The factors used initially in computing the seasonally adjusted indexes were derived by the BLS Seasonal Factor Method using data for 1956–65. These factors are updated at the end of each calendar year.

The seasonal adjustment does not affect the procedure for computing the original indexes. The unadjusted all items and group indexes are derived as described above. The seasonal calculations are a separate operation designed to make available data from which normal seasonal fluctuations have been removed to facilitate analysis.

LIMITATIONS OF THE INDEX

The Consumer Price Index is not an exact measurement of price changes. It is subject to sampling errors which cause it to deviate somewhat from the results that would be obtained if actual records of all retail purchases by wage earners and clerical workers could be used to compile the index. These estimating or sampling errors are not mistakes in the index calculation. They are unavoidable. They could be reduced by using much larger samples, but the cost is prohibitive. Furthermore, the index is believed to be sufficiently accurate for most of the practical uses made of it.

Another kind of error occurs because people who give information do not always report accurately. The Bureau makes every effort to keep these errors to a minimum, and corrects them whenever they are discovered subsequently. Precautions are taken to guard against errors in pricing, which would affect the index most seriously. The field representatives who collect the price data and the commodity specialists and clerks who process them are well trained to watch for unusual deviations in prices that might be due to errors in reporting.

The Consumer Price Index represents the average movement of prices for wage earners and clerical workers as a broad group, but not necessarily the change in prices paid by any one family or small group of families. The index is not directly applicable to any other occupational group. Some families may find their outlays changing because of changes in factors other than prices, such as family composition. The index measures only the change in prices and none of the other factors that affect family living expenses.

In many instances, changes in quoted prices are accompanied by changes in the quality of consumer goods and services. Also, new products are introduced frequently, which bear little resemblance to products previously on the market; hence, direct price comparisons cannot be made. The Bureau of Labor Statistics makes every effort to adjust quoted prices for changes in quality, and has developed special procedures for this purpose, including the use of technical specifications and highly trained personnel referred to previously. Nevertheless, some residual effects of quality changes on quoted prices undoubtedly do affect the movement of the Consumer Price Index either downward or upward from time to time.

Appendix: Explanation of the Index Formula

In the absence of major weight revisions or sample changes, the index formula is most simply expressed as:

$$I_{i:o} = \frac{\Sigma\,(q_o p_i)}{\Sigma\,(q_o p_o)} \times 100 = \frac{\Sigma\,(q_o p_o)\left(\dfrac{p_i}{p_o}\right)}{\Sigma\,(q_o p_o)} \times 100 \qquad (1)$$

This is the customary, oversimplified way of writing a price index formula to show that the q's are held constant between major revisions. In actual practice, the basic data for weights are values, and the quantity and price elements of the "pq" values (p's and q's) are not separated.

With a weight revision, the formula becomes

$$I_{i:o} = \frac{\sum (q_o p_{i-s})}{\sum (q_o p_o)} \times \frac{\sum (q_a p_i')}{\sum (q_a p_{i-s}')} \times 100 \qquad (2)$$

where q is a derived composite of the annual quantities purchased in a weight base period for a bundle of goods and services to be represented by a specific item priced,

p and p' are the average prices of the specific commodities or services selected for pricing (the superscript indicates that the average prices are not necessarily derived from identical samples of outlets and specifications over long periods),

$i - s$ is the month preceding a weight revision (most recently, December 1963),

i is the current month,

a is the period of the most recent consumer expenditure survey (1960–61) from which the revised weights were derived,

o is the reference base period of the index (1957–59).

The $(q_o p_o)$ or $(q_a p'_{i\text{-}s})$ base "weights" for a given priced item are the average expenditures in a weight base period represented by that item (including expenditures for the item itself and for other similar nonpriced items).

In actual practice, this expenditure is projected forward for each pricing period by the price relative for the priced item:

$$(q_a p_i) = (q_a p_{i-1}) \left(\frac{p_i}{p_{i-1}} \right)$$

In practice, then, the index formula is as follows:

$$I_{i:o} = \frac{\sum (q_o p_{i-s})}{\sum (q_o p_o)} \times \frac{\sum (q_a p_{i-1}')}{\sum (q_a p_{i-s}')} \times \frac{\sum (q_a p_{i-1}') \left(\frac{p_i'}{p_{i-1}'} \right)}{\sum (q_a p_{i-1}')} \times 100 \quad (3)$$

Thus, although the cost weight changes with every change in price, the implicit quantity (q_o) or (q_a) remains fixed between major weight revisions.

The long-term price relative for each priced item p_i in reality is

$$R_{i:o} = \left(\frac{p_1}{p_o}\right) \cdot \left(\frac{p_2'}{p_1'}\right) \cdot \left(\frac{p_3''}{p_2''}\right) \cdot \quad \cdots\cdots\cdots \quad \cdot \left(\frac{p_i'''' \quad \cdots\cdots\cdots}{p_{i-1}'''' \quad \cdots\cdots\cdots}\right)$$

That is, $R_{i:o}$ is the product of a number of short-term relatives. The superscripts on the p's indicate that these average prices are not necessarily derived from identical samples of outlets and specifications over long periods. This chaining of monthly, or quarterly, price relatives based on comparable specifications in successive periods allows the requisite flexibility to make substitutions of items, specifications, and outlets.

PROBABILITY
AND
STATISTICAL
INFERENCE

A probability is a measure of the likelihood of occurrence of an event. The theory of probability is the base on which the field of statistics rests, and it is impossible to understand statistics without having at least a minimum knowledge of probability theory. The opening paper by Richard von Mises is a classic statement of the objectivist view of probability theory. According to von Mises, the probability of an attribute is "the relative frequency of the observed attribute . . . if the observations were indefinitely continued." For example, if we toss a coin repeatedly, and if we compute after each toss the proportion of times it has come up heads, the probability of its coming up heads is the value of this proportion after a very large number of tosses.

In the following article, R. A. Fisher, one of the great figures of modern statistics, shows how probability theory can be used for purposes of statistical inference. The broad task of statistical inference is to provide measures of the uncertainty of conclusions drawn from data. As a simple example, Fisher takes the case of a lady who declares that she can tell by taste whether the milk or sugar was first added to a cup of tea. He considers how an experiment might be designed to test this assertion and, in the course of the discussion, takes up many important aspects of the theory of hypothesis testing.

Recent years have witnessed a revolution in statistics. Many leading statisticians have adopted an approach to probability theory and statistical inference that differs sharply from that represented by von Mises and Fisher. Although much of this approach has been developed and elaborated since World War II, its origins go back to an 18th-century scholar, Thomas Bayes. In the next paper, Jack Hirshleifer compares the newer Bayesian approach with the classical approach. As he points out, the "crux of this statistical revolution is the explicit use of a priori information, in the form of a 'subjective' probability distribution for the unknown parameter under investigation. The subjective probability distribution describes the decision-maker's state of information or degree of belief as to the several different conceivable values that the unknown parameter may take."

A leader of this revolution is L. J. Savage, who discusses in the final paper in Part Two the reasons why he believes that personal, or subjective, probability is the only probability concept that is essential to science or decision-making. He begins by answering some of the objections to the personalistic view. Then he takes up some of the difficulties in the objectivist views advanced by von Mises, Fisher, and others. He argues that "objectivist views, by their nature, must in principle regard decision as secondary to probability, if relevant at all," that they attach probabilities only to very special events, and that they can be charged with circularity. The Bayesian revolution, led by Savage, Raiffa, Schlaifer, and others, has had a very great impact on statistics, although the classical approach still has many adherents.

Probability: An Objectivist View

RICHARD VON MISES

Richard von Mises is generally regarded as one of the ablest defenders of an objective view of probability. This piece comes from his classic book, *Probability, Statistics and Truth,* published in 1939.

To illustrate the apparent contrast between statistics and truth, may I quote a remark I once overheard: "There are three kinds of lies: white lies, which are justifiable; common lies—these have no justification; and statistics." Our meaning is similar when we say: "Anything can be proved by figures"; or, modifying a well-known quotation from Goethe, with numbers "all men may contend their charming systems to defend."

At the basis of all such remarks lies the conviction that conclusions drawn from statistical considerations are at best uncertain and at worst misleading. I do not deny that a great deal of meaningless and unfounded talk is presented to the public in the name of statistics. But my purpose is to show that, starting from statistical observations and applying to them a clear and precise concept of probability, it is possible to arrive at conclusions that are just as reliable and 'truth-full' and quite as practically useful as those obtained in any other exact science.

LIMITATION OF SCOPE

From the complex of ideas that are colloquially covered by the word "probability," we must remove all those that remain outside the theory we are endeavoring to formulate. I shall therefore begin with a preliminary delimitation of our concept of probability; this will be developed into a more precise definition during the course of our discussion.

Our probability theory has nothing to do with questions such as: "Is there a probability of Germany's being at some time in the future involved in a war with Liberia?" Again, the question of the "probability" of the correct interpretation of a certain passage from the *Annals* of Tacitus has nothing in common with our theory. It need hardly be pointed out that we are likewise unconcerned with the "intrinsic probability" of a work of art. The relation of our theory to Goethe's superb dialogue on *Truth and Probability in Fine Art* is thus only one of similarity in the sounds of words and consequently is irrelevant. We shall not deal with the problem of the historical accuracy of Biblical narratives, although it is interesting to note that a Russian mathematician, A. Markoff, inspired by the ideas of the 18th century Enlightenment, wished to see the theory of probability applied to this subject. Similarly, we shall not concern ourselves with any of those problems of the moral sciences that were so ingeniously treated by Laplace in his *Essai Philosophique*. The unlimited extension of the validity of the exact sciences was a characteristic feature of the exaggerated rationalism of the 18th century. We do not intend to commit the same mistake.

Problems such as the probable reliability of witnesses and the correctness of judicial verdicts lie more or less on the boundary of the region that we are going to include in our treatment. These problems have been the subject of many scientific discussions; Poisson chose them as the title of his famous book.

To reach the essence of the problems of probability, we must consider, for example, the probability of winning in a carefully defined game of chance. Is it sensible to bet that a "double 6" will appear at least once if 2 dice are thrown 24 times? Is this result "probable"? More exactly, how great is its probability? Such are the questions we feel able to answer. Many problems of considerable importance in everyday life belong to the same class and can be treated in the same way; examples of these are many problems connected with insurance, such as those concerning the probability of illness or death occurring under carefully specified conditions, the premium that must be asked for insurance against a particular kind of risk, and so forth.

Besides the games of chance and certain problems relating to social mass phenomena, there is a third field in which our concept has a useful application. This is in the treatment of certain mechanical and physical phenomena. Typical examples may be seen in the movement of molecules in a gas or in the random motion of colloidal particles which can be observed with the ultramicroscope. ("Colloid" is the name given to a system of very fine particles freely suspended in a medium, with the size of the particles so minute that the whole appears to the naked eye to be a homogeneous liquid.)

UNLIMITED REPETITION

What is the common feature in the last three examples and what is the essential distinction between the meaning of "probability" in these cases and its meaning in the earlier examples which we have excluded from our treatment? One common feature can be recognized easily, and we think it crucial. In games of chance, in the problems of insurance, and in the molecular processes, we find events repeating themselves again and again. They are mass phenomena or repetitive events. The throwing of a pair of dice is an event that can theoretically be repeated an unlimited number of times, for we do not take into account the wear of the box or the possibility that the dice may break. If we are dealing with a typical problem of insurance, we can imagine a great army of individuals insuring themselves against the same risk, and the repeated occurrence of events of a similar kind (e.g., deaths) are registered in the records of insurance companies. In the third case, that of the molecules or colloidal particles, the immense number of particles partaking in each process is a fundamental feature of the whole conception.

On the other hand, this unlimited repetition, this "mass character," is typically absent in the case of all the examples previously excluded. The implication of Germany in a war with the Republic of Liberia is not a situation which frequently repeats itself; the uncertainties that occur in the transcription of ancient authors are, in general, of a too individual character for them to be treated as mass phenomena. The question of the trustworthiness of the historical narratives of the Bible is clearly unique and cannot be considered as a link in a chain of analogous problems. We classified the reliability and trustworthness of witnesses and judges as a borderline case since we may feel reasonable doubt whether similar situations occur sufficiently frequently and uniformly for them to be considered as repetitive phenomena.

We state here explicitly: The rational concept of probability, which is the only basis of probability calculus, applies only to problems in which either the same event repeats itself again and again, or a great number of uniform elements is involved at the same time. Using the language of physics, we may say that in order to apply the theory of probability we must have a practically unlimited sequence of uniform observations.

THE COLLECTIVE

A good example of a mass phenomenon suitable for the application of the theory of probability is the inheritance of certain characteristics, e.g., the color of flowers resulting from the cultivation of large numbers of plants of a given species from a given seed. Here we can easily recognize what is meant by the words "a repetitive event." There is primarily a single instance: the growing of one plant and the observation of the color of its flowers. Then comes the comprehensive treatment of a great number of such instances, considered as parts of one greater unity. The individual elements belonging to this unity differ from each other only with respect to a single attribute, the color of the flowers.

In games of dice, the individual event is a single throw of the dice from the box and the attribute is the observation of the number of points shown by the dice. In the game "heads or tails," each toss of the coin is an individual event, and the side of the coin that is uppermost is the attribute. In life insurance the single event is the life of the individual and the attribute observed is either the age at which the individual dies or, more generally, the moment at which the insurance company becomes liable for payment. When we speak of "the probability of death," the exact meaning of this expression can be defined in the following way only. We must not think of an individual, but of a certain class as a whole, "all insured men 41 years old living in a given country and not engaged in certain dangerous occupations." A probability of death is attached to this class of men or to another class that can be defined in a similar way. We can say nothing about the probability of death of an individual even if we know his condition of life and health in detail. The phrase "probability of death," when it refers to a single person, has no meaning at all for us. This is one of the most important consequences of our definition of probability, and we shall discuss this point in greater detail later on.

We must now introduce a new term, which will be very useful during the future course of our argument. This term is "the collective," and it denotes a sequence of uniform events or processes that differ by certain observable attributes, say colors, numbers, or anything else. In a preliminary way we state: All the peas grown by a botanist concerned with the problem of heredity may be considered as a collective, the attributes in which we are interested being the different colors of the flowers. All the throws of dice made in the course of a game form a collective wherein the attribute of the single event is the number of points thrown. Again, all the molecules in a given volume of gas may be considered as a collective, and the attribute of a single molecule might be its velocity. A further example of a collective is the whole class of insured men and women whose ages at death have been registered by an insurance office. The principle which underlies the whole of our treatment of the probability problem is that a collective must exist before we begin to speak of probability. The definition of probability which we shall give is only concerned with "the probability of encountering a certain attribute in a given collective."

THE FIRST STEP TOWARD A DEFINITION

After our previous discussion it should not be difficult to arrive at a rough form of definition of probability. We may consider a game with two dice. The attribute of a single throw is the sum of the points showing on the upper sides of the two dice. What shall we call the probability of the attribute "12," that is, the case of each die showing six points? When we have thrown the dice a large number of times, say 200, and noted the results, we find that 12 has appeared a certain number of times, perhaps five times. The ratio $5/200 = 1/40$ is called the frequency, or more accurately the relative frequency, of the attribute "12" in the first 200 throws. If we continue the game for another 200 throws we can find the corresponding relative frequency for 400 throws, and so on. The ratios that are obtained in this way will differ a little from the first one, 1/40. If the ratios were to continue to show considerable variation after the game had been repeated 2000, 4000, or a still larger number of times, then the question whether there is a definite probability of the result "12" would not arise at all. It is essential for the theory of probability that experience has shown that in the game of dice, as in all the other mass phenomena which we have mentioned, the relative frequencies of certain attributes become more and more stable as the number of observations is increased. We shall discuss the idea of "the limiting value of the relative frequency" later on; meanwhile, we assume that the frequency is being computed with a limited accuracy only, so that small deviations are not perceptible. This approximate value of the relative frequency we shall, preliminarily, regard as the probability of the attribute in question, e.g., the probability of the result "12" in the game of dice. It is obvious that, if we define probability in this way, it will be a number less than one, that is, a proper fraction.

TWO DIFFERENT PAIRS OF DICE

I have here two pairs of dice that are apparently alike. By repeatedly throwing one pair, it is found that the relative frequency of the "double 6" approaches a value of 0.028, or 1/36, as the number of trials is increased. The second pair shows a relative frequency for the "12," which is four times as large. The first pair is usually called a pair of true dice, the second is called biased, but our definition of probability applies equally to both pairs. Whether or not a die is biased is as irrelevant for our theory as is the moral integrity of a patient when a physician is diagnosing his illness. Eighteen hundred throws were made with each pair of these dice. The sum "12" appeared 48 times with the first pair and 178 times with the second. The relative frequencies are

$$\frac{48}{1800} = \frac{1}{37.5} = 0.027$$

and

$$\frac{178}{1800} = \frac{1}{10.1} = 0.099$$

These ratios became practically constant toward the end of the series of trials. For instance, after the 1500th throw they were 0.023 and 0.094, respectively. The differences between the values calculated at this stage and later on did not exceed 10 to 15 percent.

It is impossible for me to show you a lengthy experiment in the throwing of dice during the course of this lecture, since it would take too long. It is sufficient to make a few trials with the second pair of dice to see that at least one 6 appears at nearly every throw; this is a result very different from that obtained with the other pair. In fact, it can be shown that, if we throw one of the dice belonging to the second pair, the relative frequency with which a single 6 appears is about 1/3, whereas for either of the first pair this frequency is almost exactly 1/6. In order to realize clearly what our meaning of probability implies, it will be useful to think of these two pairs of dice as often as possible; each pair has a characteristic probability of showing "double 6," but these probabilities differ widely.

Here we have the "primary phenomenon" (Urphän omen) of the theory of probability in its simplest form. The probability of a 6 is a physical property of a given die and is a property analogous to its mass, specific heat, or electrical resistance. Similarly, for a given *pair of dice* (including, of course, the total setup) the probability of a "double 6" is a characteristic property, a physical constant belonging to the experiment as a whole and comparable with all its other physical properties. The theory of probability is only concerned with relations existing between physical quantities of this kind.

LIMITING VALUE OF RELATIVE FREQUENCY

I have used the expression "limiting value," which belongs to higher analysis, without further explanation. We do not need to know much about the mathematical definition of this expression, since we propose to use it in a manner that can be understood by anyone, however ignorant of higher mathematics. Let us calculate the relative frequency of an attribute in a collective. This is the ratio of the number of cases in which the attribute has been found to the total number of observations. We shall calculate it with a certain limited accuracy, i.e., to a certain number of decimal places without asking what the following figures might be. Suppose, for instance, that we play "heads or tails" a number of times and calculate the relative frequency of "heads." If the number of games is increased and if we always stop at the same decimal place in calculating the relative frequency, then, eventually, the results of such calculations will cease to change. If the relative frequency of heads is calculated accurately to the first decimal place, it would not be difficult to attain constancy in this first approximation. In fact, perhaps after some 500 games, this first approximation will reach the value of 0.5 and will not change afterwards. It will take us much longer to arrive at a constant value for the second approximation, calculated to two decimal places. For this purpose it may be necessary to calculate the relative frequency in intervals of, say, 500 casts, i.e., after the 500th, 1000th, 1500th, and 2000th cast, and so on. Perhaps more than 10,000 casts will be required to show that now the second figure also ceases to change and remains equal to 0, so that the relative frequency remains constantly 0.50. Of course it is impossible to continue an experiment of this kind indefinitely. Two experimenters, cooperating efficiently, may be able to make up to 1000 observations per hour, but not more. Imagine, for example, that the experiment has been continued for 10 hours and that the relative frequency remained constant at 0.50 during the last two hours. An astute observer might perhaps have managed to calculate the third figure as well, and might have found that the changes in this figure during the last hours, although still occurring, were limited to a comparatively narrow range.

Considering these results, a scientifically trained mind may easily accept the hypothesis that by continuing this play for a sufficiently long time under conditions that do not change (insofar as this is practically possible), one would arrive at constant values for the third, fourth, and all the following decimal places as well. The expression we used, stating that the relative frequency of the attribute "heads" tends to a limit, is no more than a short description of the situation assumed in this hypothesis.

Take a sheet of graph paper and draw a curve with the total number of observations as abscissas and the value of the relative frequency of the result "heads" as ordinates. At the beginning this curve shows large oscillations, but gradually they become smaller and smaller, and the curve approaches a straight horizontal line. At last the oscillations become so small that they cannot be represented on the diagram, even if a very large scale is used. It is of no importance for our purpose if the ordinate of the final horizontal line is 0.6, or any other value, instead of 0.5. The important point is the existence of this straight line. The ordinate of this horizontal line is the limiting value of the relative frequency represented by the diagram, in our case the relative frequency of the event "heads."

Let us now add further precision to our previous definition of the collective. We will say that a collective is a mass phenomenon or a repetitive event, or, simply, a long sequence of observations for which there are sufficient reasons to believe that the relative frequency of the observed attribute would tend to a fixed limit if the observations were indefinitely continued. This limit will be called *the probability of the attribute considered within the given collective*. This expression being a little cumbersome, it is obviously not necessary to repeat it always. Occasionally, we may speak simply of the *probability of "heads."* The important thing to remember is that this is only an abbreviation, and that we should know exactly the kind of collective to which we are referring. "The probability of winning a battle," for instance, has no place in our theory of probability, because we cannot think of a collective to which it belongs. The theory of probability cannot be applied to this problem any more than the physical concept of work can be applied to the calculation of the "work" done by an actor in reciting his part in a play.

Statistical Inference

RONALD A. FISHER

Ronald A. Fisher was one of the great figures of modern statistics. This article comes from his book, *The Design of Experiments,* published in 1949.

1. THE GROUNDS ON WHICH EVIDENCE IS DISPUTED

WHEN any scientific conclusion is supposed to be proved on experimental evidence, critics who still refuse to accept the conclusion are accustomed to take one of two lines of attack. They may claim that the *interpretation* of the experiment is faulty, that the results reported are not in fact those which should have been expected had the conclusion drawn been justified, or that they might equally well have arisen had the conclusion drawn been false. Such criticisms of interpretation are usually treated as falling within the domain of *statistics.* They are often made by professed statisticians against the work of others whom they regard as ignorant of or incompetent in statistical technique; and, since the interpretation of any considerable body of data is likely to involve computations, it is natural enough that questions involving the logical implications of the results of the arithmetical processes employed should be relegated to the statistician. At least I make no complaint of this convention. The statistician cannot evade the responsibility for understanding the processes he applies or recommends. My immediate point is that the questions involved can be dissociated from all that is strictly technical in the statistician's craft, and, *when so detached,* are questions only of the right use of human reasoning powers, with which all intelligent people, who hope to be intelligible, are equally concerned, and on which the statistician, as such, speaks with no special authority. The statistician cannot excuse himself from the duty of getting his head clear on the principles of scientific inference, but equally no other thinking man can avoid a like obligation.

The other type of criticism to which experimental results are exposed is that the experiment itself was ill designed, or, of course, badly executed. If we suppose that the experimenter did what he intended to do, both of these points come down to the question of the *design*, or the *logical structure* of the experiment. This type of criticism is usually made by what I might call a heavyweight *authority*. Prolonged experience, or at least the long possession of a scientific reputation, is almost a prerequisite for developing successfully this line of attack. Technical details are seldom in evidence. The authoritative assertion "His *controls* are *totally* inadequate" must have temporarily discredited many a promising line of work; and such an authoritarian method of judgment must surely continue, human nature being what it is, so long as theoretical notions of the principles of experimental design are lacking—notions just as clear and explicit as we are accustomed to apply to technical details.

Now the essential point is that the two sorts of criticism I have mentioned are aimed only at different aspects of the same whole, although they are usually delivered by different sorts of people and in very different language. If the design of an experiment is faulty, any method of interpretation that makes it out to be decisive must be faulty too. It is true that there are a great many experimental procedures which are well designed in that they *may* lead to decisive conclusions, but on other occasions may fail to do so; in such cases, if decisive conclusions are in fact drawn when they are unjustified, we may say that the fault is wholly in the interpretation, not in the design. But the fault of interpretation, even in these cases, lies in overlooking the characteristic features of the design which lead to the results being sometimes inconclusive, or conclusive on some questions but not on all. To understand correctly the one aspect of the problem is to understand the other. Statistical procedure and experimental design are only two different aspects of the same whole, and that whole comprises all the logical requirements of the complete process of adding to natural knowledge by experimentation.

2. THE MATHEMATICAL ATTITUDE TOWARD INDUCTION

In the foregoing paragraphs the subject matter of this book has been regarded from the point of view of an experimenter, who wishes to carry out his work competently, and having done so wishes to safeguard his results, so far as they are validly established, from ignorant criticism by different sorts of superior persons. I have assumed, as the experimenter always does assume, that it *is* possible to draw valid inferences from the results of experimentation; that it is possible to argue from consequences to causes, from observations to hypotheses; as a statistician would say, from a sample to the population from which the sample was drawn, or, as a logician might put it, from the particular to the general. It is, however, certain that many mathematicians, if pressed on the point, would say that it is not possible rigorously to argue from the particular to the general; that all such arguments must involve some sort of guesswork, which they might admit to be plausible guesswork, but the rationale of which, they would be unwilling, as mathematicians, to discuss. We may at once admit that any inference from the particular to the general must be attended with some degree of uncertainty, but this is not the same as to admit that such inference cannot be absolutely rigorous, for the nature and degree of the uncertainty may itself be capable of rigorous expression. In the theory of probability, as developed in its application to games of chance, we have the classic example proving this possibility. If the gamblers' apparatus is really *true* or unbiased, the probabilities of the different possible events, or combinations of events, can be inferred by a rigorous deductive argument, although the outcome of any particular game is recognized to be uncertain. The mere fact that inductive inferences are uncertain cannot, therefore, be accepted as precluding perfectly rigorous and unequivocal inference.

Naturally, writers on probability have made determined efforts to include the problem of inductive inference within the ambit of the theory of mathematical probability, developed in discussing deductive problems arising in games of chance. To illustrate how much was at one time thought to have been achieved in this way, I may quote a very lucid statement by Augustus de Morgan, published in 1838, in the preface to his essay on probabilities in *The Cabinet Cyclopædia*. At this period confidence in the theory of inverse probability, as it was called, had reached, under the influence of Laplace, its highest point. Boole's criticisms had not yet been made, nor the more decided rejection of the theory by Venn, Chrystal, and later writers. De Morgan is speaking of the advances in the theory which were leading to its wider application to practical problems.

"There was also another circumstance that stood in the way of the first investigators, namely, the not having considered, or, at least, not having discovered the method of reasoning from the happening of an event to the probability of one or another cause. Given an hypothesis presenting the necessity of one or another out of a certain, and not very large, number of consequences, they could determine the chance that any given one or other of those consequences should arrive; but given an event as having happened, and which might have been the consequence of either of several different causes or explicable by either of several different hypotheses, they could not infer the probability with which the happening of the event should cause the different hypotheses to be viewed. But just as in natural philosophy the selection of an hypothesis by means of observed facts is always preliminary to any attempt at deductive discovery; so in the application of the notion of probability to the actual affairs of life, the process of reasoning from observed events to their most probable antecedents must go before the direct use of any such antecedent, cause, hypothesis, or whatever it may be correctly termed. These two obstacles, therefore, the mathematical difficulty, and the want of an inverse method, prevented the science from extending its views beyond problems of that simple nature which games of chance present."

Referring to the inverse method, he later adds: "This was first used by the Rev. T. Bayes, and the author, though now almost forgotten, deserves the most honorable remembrance from all who treat the history of this science."

3.   THE REJECTION OF INVERSE PROBABILITY

Whatever may have been true in 1838, it is certainly not true today that Thomas Bayes is almost forgotten. That he seems to have been the first man in Europe to have seen the importance of developing an exact and quantitative theory of inductive reasoning, of arguing from observational facts to the theories that might explain them, is surely a sufficient claim to a place in the history of science. But he deserves honorable remembrance for one fact, also, in addition to those mentioned by de Morgan. Having perceived the problem and devised an axiom which, if its truth were granted, would bring inverse inferences within the scope of the theory of mathematical probability, he was sufficiently critical of its validity to withhold his entire treatise from publication until his doubts should have been satisfied. In the event, the work was published after his death by his friend, Price, and we cannot say what views he ultimately held on the subject.

The discrepancy of opinion among historical writers on probability is so great that to mention the subject is unavoidable. It would, however, be out of place here to argue the point in detail. I will only state three considerations which will explain why, in the practical applications of the subject, I shall not assume the truth of Bayes' axiom. Two of these reasons would, I think, be generally admitted, but the first, I can well imagine, might be indignantly repudiated in some quarters. The first is this: The axiom leads to apparent mathematical contradictions. In explaining these contradictions away, advocates of inverse probability seem forced to regard mathematical probability, not as an objective quantity measured by observed frequencies, but as measuring merely psychological tendencies, theorems respecting which are useless for scientific purposes.

My second reason is that it is the nature of an axiom that its truth should be apparent to any rational mind that fully apprehends its meaning. The axiom of Bayes has certainly been fully apprehended by a good many rational minds, including that of its author, without carrying this conviction of necessary truth.

This, alone, shows that it cannot be accepted as the axiomatic basis of a rigorous argument.

My third reason is that inverse probability has been only very rarely used in the justification of conclusions from experimental facts, although the theory has been widely taught, and is widespread in the literature of probability. Whatever the reasons are which give experimenters confidence that they can draw valid conclusions from their results, they seem to act just as powerfully whether the experimenter has heard of the theory of inverse probability or not.

4. THE LOGIC OF THE LABORATORY

In fact, I propose to consider a number of different types of experimentation, with especial reference to their logical structure, and to show that when the appropriate precautions are taken to make this structure complete, entirely valid inferences may be drawn from them, without using the disputed axiom. *If* this can be done, we shall, in the course of studies having directly practical aims, have overcome the theoretical difficulty of inductive inferences.

5. STATEMENT OF EXPERIMENT

A lady declares that by tasting a cup of tea made with milk she can discriminate whether the milk or the tea infusion was first added to the cup. We will consider the problem of designing an experiment by means of which this assertion can be tested. For this purpose let us first lay down a simple form of experiment with a view to studying its limitations and its characteristics, both those that appear to be essential to the experimental method, when well developed, and those that are not essential but auxiliary.

Our experiment consists in mixing eight cups of tea, four in one way and four in the other, and presenting them to the subject for judgment in a random order. The subject has been told in advance of what the test will consist, namely that she will be asked to taste eight cups, that these shall be four of each kind, and that they shall be presented to her in a random order, that is, in an order not determined arbitrarily by human choice, but by the actual manipulation of the physical apparatus used in games of chance, cards, dice, roulettes, etc., or, more expeditiously, from a published collection of random sampling numbers purporting to give the actual results of such manipulation. Her task is to divide the eight cups into two sets of four, agreeing, if possible, with the treatments received.

6.   INTERPRETATION AND ITS REASONED BASIS

In considering the appropriateness of any proposed experimental design, it is always needful to forecast all possible results of the experiment, and to have decided without ambiguity what interpretation shall be placed upon each one of them. Further, we must know by what argument this interpretation is to be sustained. In the present instance we may argue as follows. There are 70 ways of choosing a group of 4 objects out of 8. This may be demonstrated by an argument familiar to students of "permutations and combinations," namely, that if we were to choose the 4 objects in succession we should have successively 8, 7, 6, 5 objects to choose from, and could make our succession of choices in $8 \times 7 \times 6 \times 5$, or 1680 ways. But in doing this we have not only chosen every possible set of 4, but every possible set in every possible order; and since 4 objects can be arranged in order in $4 \times 3 \times 2 \times 1$, or 24 ways, we may find the number of possible choices by dividing 1680 by 24. The result, 70, is essential to our interpretation of the experiment. At best the subject can judge rightly with every cup and, knowing that 4 are of each kind, this amounts to choosing, out of the 70 sets of 4 that might be chosen, that particular one which is correct. A subject without any faculty of discrimination would in fact divide the 8 cups correctly into two sets of 4 in one trial out of 70, or, more properly, with a frequency which would approach 1 in 70 more and more nearly the more often the test were repeated. Evidently this frequency, with which unfailing success would be achieved by a person lacking altogether the faculty under test, is calculable from the number of cups used. The odds could be made much higher by enlarging the experiment, while, if the experiment were much smaller even the greatest possible success would give odds so low that the result might, with considerable probability, be ascribed to chance.

7. THE TEST OF SIGNIFICANCE

It is open to the experimenter to be more or less exacting in respect of the smallness of the probability he would require before he would be willing to admit that his observations have demonstrated a positive result. It is obvious that an experiment would be useless of which no possible result would satisfy him. Thus, if he wishes to ignore results having probabilities as high as 1 in 20—the probabilities being of course reckoned from the hypothesis that the phenomenon to be demonstrated is in fact absent—then it would be useless for him to experiment with only 3 cups of tea of each kind. For 3 objects can be chosen out of 6 in only 20 ways, and therefore complete success in the test would be achieved without sensory discrimination, *i.e.*, by "pure chance," in an average of 5 trials out of 100. It is usual and convenient for experimenters to take 5 percent as a standard level of significance, in the sense that they are prepared to ignore all results that fail to reach this standard, and, by this means, to eliminate from further discussion the greater part of the fluctuations which chance causes have introduced into their experimental results. No such selection can eliminate the whole of the possible effects of chance coincidence, and if we accept this convenient convention, and agree that an event which would occur by chance only once in 70 trials is decidedly "significant," in the statistical sense, we thereby admit that no isolated experiment, however significant in itself, can suffice for the experimental demonstration of any natural phenomenon; for the "one chance in a million" will undoubtedly occur, with no less and no more than its appropriate frequency, however surprised we may be that it should occur to *us*. In order to assert that a natural phenomenon is experimentally demonstrable we need, not an isolated record, but a reliable method of procedure. In relation to the test of significance, we may say that a phenomenon is experimentally demonstrable when we know how to conduct an experiment that will rarely fail to give us a statistically significant result.

Returning to the possible results of the psychophysical experiment, having decided that if every cup were rightly classified a significant positive result would be recorded, or, in other words, that we should admit that the lady had made good her claim, what should be our conclusion if, for each kind of cup, her judgments are 3 right and 1 wrong? We may take it, in the present discussion, that any error in one set of judgments will be compensated by an error in the other, since it is known to the subject that there are four cups of each kind. In enumerating the number of ways of choosing 4 things out of 8, such that 3 are right and 1 wrong, we may note that the 3 right may be chosen, out of the 4 available, in 4 ways and, independently of this choice, that the 1 wrong may be chosen, out of the 4 available, also in 4 ways. So that in all we could make a selection of the kind supposed in 16 different ways. A similar argument shows that, in each kind of judgment, 2 may be right and 2 wrong in 36 ways, 1 right and 3 wrong in 16 ways and none right and 4 wrong in 1 way only. It should be noted that the frequencies of these five possible results of the experiment make up together, as it is obvious they should, the 70 cases out of 70.

It is obvious, too, that 3 successes to 1 failure, although showing a bias, or deviation, in the right direction, could not be judged as statistically significant evidence of a real sensory discrimination. For its frequency of chance occurrence is 16 in 70, or more than 20 percent. Moreover, it is not the best possible result, and in judging of its significance we must take account not only of its own frequency, but also of the frequency of any better result. In the present instance "3 right and 1 wrong" occurs 16 times, and "4 right" occurs once in 70 trials, making 17 cases out of 70 as good as or better than that observed. The reason for including cases better than that observed becomes obvious on considering what our conclusions would have been had the case of 3 right and 1 wrong only 1 chance, and the case of 4 right 16 chances of occurrence out of 70. The rare case of 3 right and 1 wrong could not be judged significant merely because it was rare, seeing that a higher degree of success would frequently have been scored by mere chance.

8. THE NULL HYPOTHESIS

Our examination of the possible results of the experiment has therefore led us to a statistical test of significance, by which these results are divided into two classes with opposed interpretations. Tests of significance are of many different kinds, which need not be considered here. Here we are only concerned with the fact that the easy calculation in permutations which we encountered, and which gave us our test of significance, stands for something present in every possible experimental arrangement; or, at least, for something required in its interpretation. The two classes of results that are distinguished by our test of significance are, on the one hand, those which show a significant discrepancy from a certain hypothesis; namely, in this case, the hypothesis that the judgments given are in no way influenced by the order in which the ingredients have been added; and on the other hand, results that show no significant discrepancy from this hypothesis. This hypothesis, which may or may not be impugned by the result of an experiment, is again characteristic of all experimentation. Much confusion would often be avoided if it were explicitly formulated when the experiment is designed. In relation to any experiment we may speak of this hypothesis as the "null hypothesis," and it should be noted that the null hypothesis is never proved or established, but is possibly disproved, in the course of experimentation. Every experiment may be said to exist only in order to give the facts a chance of disproving the null hypothesis.

It might be argued that, if an experiment can disprove the hypothesis that the subject possesses no sensory discrimination between two different sorts of object, it must therefore be able to prove the opposite hypothesis, that she can make some such discrimination. But this last hypothesis, however reasonable or true it may be, is ineligible as a null hypothesis to be tested by experiment, because it is inexact. If it were asserted that the subject would never be wrong in her judgments, we should again have an exact hypothesis, and it is easy to see that this hypothesis could be disproved by a single failure, but could never be proved by any finite amount of experimentation. It is evident that the null hypothesis must be exact, that is, free from vagueness and ambiguity, because it must supply the basis of the "problem of distribution," of which the test of sgnificance is the solution. A null hypothesis may, indeed, contain arbitrary elements, and in more complicated cases often does so: as, for example, if it should assert that the death rates of two groups of animals are equal, without specifying what these death rates actually are. In such cases it is evidently the equality rather than any particular values of the death rates that the experiment is designed to test, and possibly to disprove.

In cases involving statistical "estimation," these ideas may be extended to the simultaneous consideration of a series of hypothetical possibilities. The notion of an error of the so-called "second kind," due to accepting the null hypothesis "when it is false" may then be given a meaning in reference to the quantity to be estimated. It has no meaning with respect to simple tests of significance, in which the only available expectations are those that flow from the null hypothesis' being true.

9. RANDOMIZATION; THE PHYSICAL BASIS OF THE VALIDITY OF THE TEST

We have spoken of the experiment as testing a certain null hypothesis, namely, in this case, that the subject possesses no sensory discrimination whatever of the kind claimed; we have, too, assigned as appropriate to this hypothesis a certain frequency distribution of occurrences, based on the equal frequency of the 70 possible ways of assigning 8 objects to two classes of 4 each; in other words, the frequency distribution appropriate to a classification by pure chance. We have now to examine the physical conditions of the experimental technique needed to justify the assumption that, if discrimination of the kind under test is absent, the result of the experiment will be wholly governed by the laws of chance. It is easy to see that it might well be otherwise. If all those cups made with the milk first had sugar added, while those made with the tea first had none, a very obvious difference in flavor would have been introduced which might well ensure that all those made with sugar should be classed alike. These groups might either be classified all right or all wrong, but in such a case the frequency of the critical event in which all cups are classified correctly would not be 1 in 70, but 35 in 70 trials, and the test of significance would be wholly vitiated. Errors equivalent in principle to this are very frequently incorporated in otherwise well-designed experiments.

It is no sufficient remedy to insist that "all the cups must be exactly alike" in every respect except that to be tested. For this is a totally impossible requirement in our example, and equally in all other forms of experimentation. In practice it is probable that the cups will differ perceptibly in the thickness or smoothness of their material, that the quantities of milk added to the different cups will not be exactly equal, that the strength of the infusion of tea may change between pouring the first and the last cup, and that the temperature also at which the tea is tasted will change during the course of the experiment. These are only examples of the differences probably present; it would be impossible to present an exhaustive list of such possible differences appropriate to any one kind of experiment, because the uncontrolled causes that may influence the result are always strictly innumerable. When any such cause is named, it is usually perceived that, by increased labor and expense, it could be largely eliminated. Too frequently it is assumed that such refinements constitute improvements to the experiment. Our view is that it is an essential characteristic of experimentation that it is carried out with limited resources, and an essential part of the subject of experimental design to ascertain how these should be best applied; or, in particular, to which causes of disturbance care should be given, and which *ought* to be deliberately ignored. To ascertain, too, for those that are not to be ignored, to what extent it is worthwhile to take the trouble to diminish their magnitude. For our present purpose, however, it is only necessary to recognize that, whatever degree of care and experimental skill is expended in equalizing the conditions, other than the one under test, which are liable to affect the result, this equalization must always be to a greater or less extent incomplete, and in many important practical cases will certainly be grossly defective. We are concerned, therefore, that this inequality, whether it be great or small, shall not impugn the exactitude of the frequency distribution, on the basis of which the result of the experiment is to be appraised.

10. THE EFFECTIVENESS OF RANDOMIZATION

The element in the experimental procedure which contains the essential safeguard is that the two modifications of the test beverage are to be prepared "in random order." This, in fact, is the only point in the experimental procedure in which the laws of chance, which are to be in exclusive control of our frequency distribution, have been explicitly introduced. The phrase "random order" itself, however, must be regarded as an incomplete instruction, standing as a kind of shorthand symbol for the full procedure of randomization, by which the validity of the test of significance may be guaranteed against corruption by the causes of disturbance which have not been eliminated. To demonstrate that, with satisfactory randomization, its validity is, indeed, wholly unimpaired, let us imagine all causes of disturbance—the strength of the infusion, the quantity of milk, the temperature at which it is tasted, etc.—to be predetermined for each cup; then since these, on the null hypothesis, are the only causes influencing classification, we may say that the probabilities of each of the 70 possible choices or classifications which the subject can make are also predetermined. If, now, after the disturbing causes are fixed, we assign, strictly at random, 4 out of the 8 cups to each of our experimental treatments, then every set of 4, whatever its probability of being so classified, will certainly have a probability of exactly 1 in 70 of *being* the 4, for example, to which the milk is added first. However important the causes of disturbance may be, even if they were to make it certain that one particular set of 4 should receive this classification, the probability that the 4 so classified and the 4 that ought to have been so classified should be the same, must be rigorously in accordance with our test of significance.

It is apparent, therefore, that the random choice of the objects to be treated in different ways would be a complete guarantee of the validity of the test of significance, if these treatments were the last in time of the stages in the physical history of the objects which might affect their experimental reaction. The circumstance that the experimental treatments cannot always be applied last, and may come relatively early in their history, causes no practical inconvenience; for subsequent causes of differentiation, if under the experimenter's control, as, for example, the choice of different pipettes to be used with different flasks, can either be predetermined before the treatments have been randomized, or, if this has not been done, can be randomized on their own account; and other causes of differentiation will be either (1) consequences of differences already randomized, or (2) natural consequences of the difference in treatment to be tested, of which on the null hypothesis there will be none, by definition, or (3) effects supervening by chance independently from the treatments applied. Apart, therefore, from the avoidable error of the experimenter himself introducing with his test treatments, or subsequently, other differences in treatment, the effects of which the experiment is not intended to study, it may be said that the simple precaution of randomization will suffice to guarantee the validity of the test of significance, by which the result of the experiment is to be judged.

11. THE SENSITIVENESS OF AN EXPERIMENT: EFFECTS OF ENLARGEMENT AND REPETITION

A probable objection, which the subject might well make to the experiment so far described, is that only if every cup is classified correctly will she be judged successful. A single mistake will reduce her performance below the level of significance. Her claim, however, might be, not that she could draw the distinction with invariable certainty, but that, though sometimes mistaken, she would be right more often than not; and that the experiment should be enlarged sufficiently, or repeated sufficiently often, for her to be able to demonstrate the predominance of correct classifications in spite of occasional errors.

An extension of the calculation upon which the test of significance was based shows that an experiment with 12 cups, 6 of each kind, gives, on the null hypothesis, 1 chance in 924 for complete success, and 36 chances for 5 of each kind classified right and 1 wrong. As 37 is less than a twentieth of 924, such a test could be counted as significant, although a pair of cups have been wrongly classified; and it is easy to verify that, using larger numbers still, a significant result could be obtained with a still higher proportion of errors. By increasing the size of the experiment, we can render it more sensitive, meaning by this that it will allow of the detection of a lower degree of sensory discrimination, or, in other words, of a quantitatively smaller departure from the null hypothesis. Since in every case the experiment is capable of disproving, but never of proving this hypothesis, we may say that the value of the experiment is increased whenever it permits the null hypothesis to be more readily disproved.

The same result could be achieved by repeating the experiment, as originally designed, upon a number of different occasions, counting as a success all those occasions on which 8 cups are correctly classified. The chance of success on each occasion being 1 in 70, a simple application of the theory of probability shows that 2 or more successes in 10 trials would occur, by chance, with a frequency below the standard chosen for testing significance; so that the sensory discrimination would be demonstrated, although, in 8 attempts out of 10, the subject made one or more mistakes. This procedure may be regarded as merely a second way of enlarging the experiment and, thereby, increasing its sensitiveness, since in our final calculation we take account of the aggregate of the entire series of results, whether successful or unsuccessful. It would clearly be illegitimate, and would rob our calculation of its basis, if the unsuccessful results were not all brought into the account.

12. QUALITATIVE METHODS OF INCREASING SENSITIVENESS

Instead of enlarging the experiment we may attempt to increase its sensitiveness by qualitative improvements; and these are, generally speaking, of two kinds: (1) the reorganization of its structure, and (2) refinements of technique. To illustrate a change of structure, we might consider that, instead of fixing in advance that 4 cups should be of each kind, determining by a random process how the subdivision should be effected, we might have allowed the treatment of each cup to be determined independently by chance, as by the toss of a coin, so that each treatment has an equal chance of being chosen. The chance of classifying correctly 8 cups randomized in this way, without the aid of sensory discrimination, is 1 in 2^8, or 1 in 256 chances, and there are only 8 chances of classifying 7 right and 1 wrong; consequently the sensitiveness of the experiment has been increased, while still using only 8 cups, and it is possible to score a significant success, even if one is classified wrongly. In many types of experiment, therefore, the suggested change in structure would be evidently advantageous. For the special requirements of a psychophysical experiment, however, we should probably prefer to forego this advantage, since it would occasionally occur that all the cups would be treated alike, and this, besides bewildering the subject by an unexpected occurrence, would deny her the real advantage of judging by comparison.

Another possible alteration to the structure of the experiment, which would, however, decrease its sensitiveness, would be to present determined, but unequal, numbers of the two treatments. Thus we might arrange that 5 cups should be of the one kind and 3 of the other, choosing them properly by chance, and informing the subject how many of each to expect. But since the number of ways of choosing 3 things out of 8 is only 56, there is now, on the null hypothesis, a probability of a completely correct classification of 1 in 56. It appears, in fact, that we cannot by these means do better than by presenting the two treatments in equal numbers, and the choice of this equality is now seen to be justified by its giving to the experiment its maximal sensitiveness.

With respect to the refinements of technique, we have seen above that these contribute nothing to the validity of the experiment, and of the test of significance by which we determine its result. They may, however, be important, and even essential, in permitting the phenomenon under test to manifest itself. Though the test of significance remains valid, it may be that without special precautions even a definite sensory discrimination would have little chance of scoring a significant success. If some cups were made with India and some with China tea, even though the treatments were properly randomized, the subject might not be able to discriminate the relatively small difference in flavor under investigation, when it was confused with the greater differences between leaves of different origin. Obviously, a similar difficulty could be introduced by using in some cups raw milk and in others boiled, or even condensed milk, or by adding sugar in unequal quantities. The subject has a right to claim, and it is in the interests of the sensitiveness of the experiment, that gross differences of these kinds should be excluded, and that the cups should, not as far as *possible*, but as far as is practically convenient, be made alike in all respects except that under test.

How far such experimental refinements should be carried is entirely a matter of judgment, based on experience. The validity of the experiment is not affected by them. Their sole purpose is to increase its sensitiveness, and this object can usually be achieved in many other ways, and particularly by increasing the size of the experiment. If, therefore, it is decided that the sensitiveness of the experiment should be increased, the experimenter has the choice between different methods of obtaining equivalent results; and will be wise to choose whichever method is easiest to him, irrespective of the fact that previous experimenters may have tried, and recommended as very important or even essential, various ingenious and troublesome precautions.

The Bayesian Approach to Statistical Decision

JACK HIRSHLEIFER

Jack Hirshleifer is Professor of Economics at the University of California at Los Angeles. This paper appeared in the *Journal of Business* in 1961.

1. INTRODUCTION

These notes are intended to serve as a guide to the recent ferment of ideas known generally as "the Bayesian approach" to statistical inference or decision. The discussion will presume some knowledge of the current standard or "classical" approach to the problem of inference as presented in modern elementary statistics textbooks. These new ideas, if accepted generally (and I think they ultimately will win such acceptance), will require a basic change in almost all statistical practice—at least at the relatively unsophisticated levels of "tests of significance" and "confidence-interval estimation." It is interesting that these theoretical developments have evolved in part out of the problems of *business* decision under uncertainty; in fact, Schlaifer's books—the only Bayesian texts currently available—are definitely oriented to the problem of rationalizing such business decisions.[1] Schlaifer's work cannot, in my opinion, be too highly

[1] The original book is Robert Schlaifer, *Probability and Statistics for Business Decisions* (New York: McGraw-Hill Book Co., 1959).

recommended to the student or practitioner; what he has done, almost single-handedly, is to structure into a set of operational procedures a group of revolutionary ideas which, while subverting the old order of statistical inference, had not given practitioners or consumers of statistics anything to replace the old order with. The central ideas underlying these procedures, it should be mentioned, derive primarily from the "subjectivist" or "personalist" probability theories recently expounded and developed by L. J. Savage.[2]

The crux of this statistical revolution is the explicit use of a priori information, in the form of a "subjective" probability distribution for the unknown parameter under investigation. The subjective probability distribution describes the decision-maker's state of information or degree of belief as to the several different conceivable values that the unknown parameter may take. The beliefs represented by the *prior probability distribution* are those held by the individual before the phase of the investigation under discussion; these subjective beliefs may, however, be based in part upon previous objective evidence.

To cite a simple example, suppose that a certain stake of money rests upon the outcome of a single toss of a coin. My beliefs concerning the unknown parameter of that coin (the true proportion of heads in an infinitely long sequence of tosses) might be approximated in this particular situation somewhat as follows: Suppose I think with probability 80 percent the coin is fair, but I assign a 10 percent chance to the true proportion of heads being only 0.4, and another 10 percent chance to the proportion being 0.6. In other words, I think most likely the coin is fair (or so close to fair as to make no difference); I do admit the possibility of some small degree of bias one way or the other, but there is no reason to suspect bias in one direction to be more likely than the other. In this case the use of the prior probability distribution to summarize both my degree of knowledge and my uncertainty has immediate intuitive appeal. The distribution may in turn depend upon my knowledge of the character of the individual supplying the coin (partly "subjective," partly "objective"

---

[2] L. J. Savage, *The Foundations of Statistics* (New York: John Wiley & Sons, 1954).
[A selection from the Savage volume appears after this article. *Editor*]

perhaps), possibly upon my purely subjective personal optimism or pessimism, and perhaps also upon some observations of how the coin behaved on a number of earlier occasions. However formed, the Bayesian approach requires that for rational action I must have a personal state of belief attaching a fractional probability to each possible value of the unknown parameter, prior to acting—and this is the prior distribution. The state of belief or knowledge might, of course, not be immediate and explicit in numerical form, but it could in principle be elicited by a suitable controlled experiment testing the individual's choices among various combinations of outcomes and rewards.

The *posterior probability distribution* summarizes the state of knowledge or belief of the individual after making use of the new information gained by sample evidence at the stage of the investigation under discussion. The approach as a whole is called Bayesian because of the crucial role played by Bayes's theorem in indicating how a specified prior probability distribution, when combined with sample evidence, leads to a unique posterior distribution for the unknown parameter.

The "new" statistical revolution reviewed here follows hard upon the previous "objectivist" revolution, associated primarily with the work of R. A. Fisher, and of Neyman and Pearson, and characterized by the now classical apparatus of "levels of significance" for tests of hypotheses, and "confidence coefficients" for estimates. This "old" revolution eschewed any statements about probability distributions for the unknown parameter, and attempted to arrive at procedures for coming to decisions purely on the basis of the objective evidence, given certain prespecified risks of error that the individual was willing to accept. Without going into polemics in detail here, Bayesians allege that "subjective" considerations—the intensities of prior beliefs and the economic values of making correct or incorrect decisions—enter anyway into the "objectivist" analysis by way of specification of hypotheses and of the tolerated risks of error (or significance levels). The Bayesian procedure makes the subjective elements of the decision problem explicit, bringing them into the light so that they can be carefully examined to insure consistent and logical treatment. In short, it is nonsense to assert that we can come to a decision without using *both* prior knowledge or belief (which may itself incorporate a considerable body of objective evidence previously accumulated, together with judgmental factors) *and* current objective evidence; we will do best to admit this and devise our procedures accordingly.

In fairness to the theorists expounding the "objectivist" approach, it should be mentioned that the recent development of the topic of decision theory forms a natural connection with subjectivist ideas. Indeed, some of the objectionable features of the standard approach as presented in elementary textbooks (e.g., the exclusive concentration in testing situations upon only two more or less arbitrarily selected possible values for the unknown parameter) have at least in part been remedied on the theoretical level—though not to any noticeable extent in practical applications. These notes, therefore, contrast the Bayesian approach with what is almost certainly an exaggerated or caricatured version of modern classical *thinking*, but nevertheless a fairly accurate version of current standard *practice* on the elementary level.

2.   TESTS OF HYPOTHESES

The Classical Solution

A situation in which the individual is called upon to decide between two competing hypotheses may well be regarded as the central or standard case exemplifying the modern classical approach. For reasons that will become clear later, in the Bayesian approach a simple point-estimation situation seems more central or standard, but the testing framework is most useful for illustrating the crucial differences between the two approaches. In the classical model, it is supposed that there are two competing hypotheses on which evidence will be brought to bear, the two hypotheses corresponding to a choice between only two actions. (This may be called a two-action situation.) For example, a heavily loaded plane must be granted or refused permission to take off; a lot being inspected by the purchaser must be accepted or rejected; a manufacturing process under investigation must be stopped or permitted to continue. While perhaps there are many possible values for the unknown parameter (many possible "states of the world"), a choice between only two decisions is possible. The classical textbook solution involves finding a decision rule which indicates what decision to take for each possible sample outcome. The decision rule is determined ultimately by stated risks of error, that is, values for the maximum acceptable conditional risks of making the wrong decision in one direction or the other. These conditional risks of error are measured at two specific values for the unknown parameter, one for which the first ("null") action is appropriate and one for which the second ("alternative") action is appropriate. The parameter values at which the measurements are to be taken, and the stated risks of error, are supposed to be somehow determined outside of, and prior to, the statistical analysis proper.

To provide a concrete illustration, we will imagine a sampling inspection situation. The analyst is to decide on the acceptability of a large lot (e.g., of ammunition for the Army) on the basis of the results in testing a small sample for the fraction defective. Here there are only two possible actions—accept or reject the lot —but many possible states of the world since the unknown parameter (the proportion defective, P, in the lot) can be any of a great number of discrete values between 0 and 1, inclusive. The classical approach is somewhat as follows. Let us suppose that lots of 4 percent defective or less are acceptable, and of more than 4 percent unacceptable. (The specification of the borderline value is presumably based upon economic considerations, though the question is typically left somewhat vague in textbook presentations.) If so, we establish as our null hypothesis, $H_o : P \leq .04$. This is to be tested against the alternative hypothesis, H_A, that $P > .04$. (These hypotheses are composite, however. For purposes of making calculations in terms of risks of error, it will later be necessary to specify particular parameter values within each of them.)

We must now decide on a decision rule, which involves selection of sampling method, sample size, and the critical sample outcome which divides those sample results leading to rejection of H_o from those leading to acceptance. Throughout this analysis, we will consider only the method of simple random sampling (with replacement), and for the present we will assume the sample size n fixed at 50. This leaves only the critical sample value or rejection number, p_r (the proportion of defectives in a sample of 50 which, if attained or exceeded, is to cause rejection of H_o) to be determined. The basic situation is illustrated in Fig. 1, which shows probabilities of error of two different decision rules, $p_r = .04$ and $p_r = .10$, as a function of different possible values of the unknown parameter P.[3] When $P \leqq .04$, H_o is true, so the only way we can err is in getting a sample result leading us to reject H_o. This is a Type I error. It will be noted that in this range, the rule $p_r = .04$ leads to higher probabilities of error than the rule $p_r = .10$, since, obviously, there is a higher probability of getting misleading sample results of .04 or more than of .10 or more. When $P > .04$, H_o is false, and the only way to err is in failing to reject H_o (Type II error). Here the rule p_r

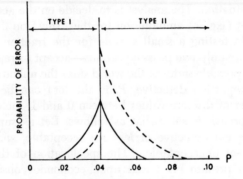

Fig. 1. Sketch of probability of error for specified decision rule, as a function of unknown parameter P.

$= .10$ leads to greater risks of error, because it is easier to get the misleading sample results below .10 than sample results below .04.

To arrive at the best decision rule under classical procedures, it is necessary to fix the maximum acceptable risks of Type I and

[3] Since the population is finite for this problem, the continuous curves drawn represent only an approximation. The true picture would show the probabilities of error as vertical bars for each discrete possible value of P. It will be noted that the risks of both kinds of error (for each decision rule) approach their maxima as P nears .04.

Type II errors for specific rather than composite null and alternative hypotheses. It is customary, in situations like this one, to fix the "level of significance" a (maximum acceptable risk of Type I error) for the borderline value $P = .04$, considered as being within H_o; β (maximum acceptable risk of Type II error) is to be fixed for some specific value within the composite H_A, but here no clear guide is given in existing presentations. Table 1 shows, for all decision rules with $n = 50$ between the rule $p_r = 0$ (i.e., always reject H_o) and the rule $p_r = .12$, the implied a at $P = .04$

TABLE 1

IMPLIED CONDITIONAL RISKS OF ERROR, a AND β, FOR DECISION RULES WITH $n = 50$

p_r	0	.02	.04	.06	.08	.10	.12
a (at $P = .04$)	1.0	.8701	.5995	.3233	.1391	.0490	.0144
β (at $P = .05$)	0	.0769	.2794	.5405	.7604	.8964	.9622
β (at $P = .06$)	0	.0453	.1900	.4162	.6473	.8206	.9224

and the implied β at two arbitrarily selected values within H_A—$P = .05$ and $P = .06$. With information like that in this table (ordinarily, either β at $P = .05$ or β at $P = .06$ would be used, not both), the analyst is supposed to be able to select his decision rule on the basis of the acceptability to him of the a and β risks of error. Without going into a detailed analysis, we may add that, if he finds these errors too great, he can reduce all his conditional risks across the board by incurring greater sampling costs—increasing his sample size, or perhaps modifying the method of sampling (going to some form of sequential sampling, for example).

From the Bayesian point of view, this procedure is defective in a number of respects. First of all, the selection of a and β is left completely up in the air, whereas we do know and can put into the analysis at least some of the considerations that should govern the selection of a and β—the economic importance of errors of the different types, and (more arguably) our prior information as to the likelihood of the different parameter values. Second, limiting the analysis to only two numerical values for the states of the world in order to get a unique a and a unique β seems highly arbitrary, even dangerous—surely it is important to consider the risk of error for *all* the possible values of *P*. In fact, practitioners following the classical analysis are likely to confuse the necessity for a choice between two *actions* or decisions, intrinsic to the 'problem, with a selection between two of the many possible states of the world—which is by no means the same thing. (We should mention here, however, that the best classical thinking does recommend "looking at" the entire risk of error picture as shown in Fig. 1. Formal computational procedures recommended in elementary textbooks, however, still involve a unique a and a unique β.) Finally, when it comes to determining sample size or method of sampling, the classical approach provides no clear procedure whereby an optimum can be obtained by balancing the costs of sampling against the gains in terms of reduction in risks of error.

We may remark that crude applications of classical techniques, especially for observational situations where sample size is fixed by the data available, generally involve deciding whether results do or do not represent "significant" divergences from the borderline value of H_o, measured exclusively in terms of an arbitrarily prespecified a, the levels commonly employed being either 5 percent or 1 percent. Table 1 illustrated how such procedures might lead to enormous risks of Type II errors. This is not to say that any classical theorists recommend neglecting Type II errors in such situations, but only that it remains common practice to do so.

THE BAYESIAN SOLUTION

The aim of the Bayesian analysis, like that of the classical analysis, can be regarded as that of establishing the optimal decision rule: selecting sampling method and size, and critical or rejection number. The Bayesian analysis makes precise and formal use of the risks of error (diagrammed in Fig. 1) of each decision rule considered as a function of the possible values for the unknown parameter (possible states of the world). Usual procedure based on classical methods throws away most of this information, employing only the risks of error for two arbitrarily or, we may say, "subjectively" selected values of the parameter within the composite H_o and H_A and thus requiring an additional arbitrary or "subjective" expression of choice among the (α, β) combinations available (as tabulated, for example, in Table 1). The Bayesian method uses all the information in Fig. 1 for each decision rule considered (and recent classical thinking would concur here); but, in addition, further information is required to calculate the best decision rule according to the Bayesian criterion: *Select the decision rule that minimizes the expected loss.*

The first additional element of information needed is the conditional loss function, where loss is to be regarded as the opportunity cost (in the economic sense) of the error. That is to say, for each possible value of the parameter (state of the world) one of the two actions will be preferable. The opportunity loss associated with choosing the preferable action is zero; however, if the inferior action has been chosen, there will be a positive opportunity cost or loss as compared with the result had the right decision been made. Figure 2 illustrates a conditional loss function for the sampling inspection problem here considered. We may note the following points:

(1) The conditional loss function is derived solely from the economics of the problem and is independent of the decision rule considered (though, of course, the probabilities of incurring these losses do depend upon the decision rule).

(2) Like the conditional risks of error, the conditional opportunity losses can be divided into Type I and Type II losses, the former applying over that range of the parameter where the null action or hypothesis is preferable or correct, and the latter where the alternative is correct.[4]

4 It is possible to dispense entirely with the terminology of null versus alternative hypotheses in the Bayesian approach; even using the classical approach, the distinction is not essential (some authors avoid it). In the classical approach, the distinction seems to apply only in selecting the specific values within the composite hypotheses at which to calculate α and β. Thus, we have seen that with H_o defined as $P \leqq .04$, α was calculated at the borderline value of $P = .04$—while with H_A defined as $P > .04$, β was calculated for some P well within this latter composite. This is the only departure from symmetry of treatment of the hypotheses in the classical method—and the Bayesian is completely symmetrical. It may be convenient to retain the term "null hypothesis," however, since statistical problems often appear as a choice between taking or not taking some positive action. For example, in a statistical quality control situation, the null hypothesis would be that the process is satisfactory so that no action is

(3) The student may think of losses as dollar values, although in certain problems it may be possible or necessary to use another "payoff" dimension (e.g., bombs on target in a military operations research problem, lives saved in a medical experiment, or "utility" in the economists' sense).

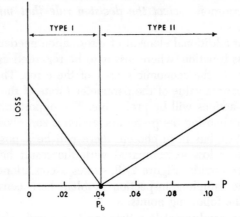

FIG. 2   Example of a loss function: loss owing to making wrong decision as a function of unknown parameter P.

(4) There will typically be some "break-even" value for the unknown parameter at which we are indifferent between the two actions; [5] in Fig. 2, this value P_b is set at .04. (Note that the

---

required—and the alternative hypothesis that something has gone wrong so that the process must be halted and inspected. An alternative interpretation is that the "null hypothesis" corresponds to the more conservative action—the decision which, if wrong, will have less drastic consequences than the other, if it should be wrong. On this interpretation, if the question arises of permitting a heavily loaded aircraft to take off, the null hypothesis would be that the plane *is* overloaded—presumably, the risk of crashing is more drastic than the economic cost of underloading an aircraft. I depart from Schlaifer in preferring the "no action" interpretation of null hypothesis to the "drastic-consequences" interpretation—the latter is difficult to define rigorously or to tie in with Bayesian ideas.

[5] This may not be strictly true when, as in this case, there are only a discrete number of possible states of the world. Thus, supposing that the lot size is 10,000, only values for P like .0399, .0400, .0401, and so forth, are possible; then it may be true that none of the possible values for P is the break-even value, one action being perhaps slightly but definitely preferable at P = .0400, and the other at .0401.

Bayesian approach makes it clear that this is not an arbitrary selection, but arises from the economics of the problem—our figure asserts that at the value of P of .3999, say, the action corresponding to H_o is preferable; at $P = .0401$, the action corresponding to H_A is preferable.)

(5) The figure illustrates a situation in which the loss due to making a wrong decision increases as P diverges from P_b, one in which it is worse to incorrectly accept a lot when its fraction defective $P = .14$ than when $P = .08$, or to reject incorrectly a good lot with $P = .01$ than a somewhat less good lot with $P = .03$.

(6) The figure shows the loss function as linear in each branch; that is only one of many possibilities.

Proponents of the classical approach would not, perhaps, deny that some such consideration of loss should enter into the determination of the specific values within H_o and H_A to be employed in calculating α and β and into the selection of the desired (α, β) combination as well; indeed, some have emphasized the concept of loss. Nevertheless, the classical approach provides no formal procedure for employing this information. However, it is on the next class of information required that the schools of thought crucially diverge; "prior probability" is the shibboleth. The classicists assert that to speak of probabilities for the unknown parameter is incorrect or meaningless, except possibly in certain very special situations. The parameter is not a random variable; any possible value considered for it either is or is not the correct one, and no probabilistic statements can be made. Bayesians reply that prior probabilities are a useful and logically consistent formalization of one's prior state of information about the unknown parameter. Users of the classical approach if they are at all reasonable will themselves take account of this information in their decisions. For example, reasonable men will insist on a higher level of significance (smaller α) before rejecting, on the basis of given sample evidence, a null hypothesis representing a strongly held belief as compared with a null hypothesis representing only a weak conjecture. By failing to formalize this information, classical analysts are in danger of making erroneous or inconsistent use of it.

Figure 3 illustrates a possible prior probability distribution [6] for the unknown parameter P.[7] We now have all the information needed to come to a Bayesian solution here, the principle being to select that decision rule minimizing expected loss, where expected loss for a given decision rule $(d.r.)$ is given by the following formula: [8]

$$EL(\text{d.r.}) = \sum_P [(\text{probability of error}|P) \times (\text{loss due to error}|P) \times (\text{prior probability of } P)]$$

[6] The probability distribution shown is continuous although, as is also true for Figs. 1 and 2, the finiteness of the population size strictly calls for a discrete representation.

[7] Students who have had the notion of "subjective" probability drilled out of them often have difficulty recapturing this rather simple and direct idea. As mentioned above, such a distribution (whether prior or posterior) represents as of that moment a formal and consistent structuring of the individual's beliefs about the unknown value of the parameter. To cite but one example, suppose the unknown parameter is a binomial proportion P of successes in a certain population. Suppose that P may take any value in the continuum between 0 and 1 (population size is infinite); a particular individual might feel that he knows for certain that $.001 < P \leq .010$, but that within this range he has no confidence of any kind in any value or any set of values over any others. This implies a uniform subjective probability distribution with the limits specified. Another individual might perhaps assign only 50 percent of probability to this range, 10 percent probability to the range below .001, and 40 percent probability to the range above .010; furthermore, within each subrange he may feel that some values are more likely than others. Operationally, we may imagine these subjective probabilities as being measured by the choices an individual makes when faced with certain betting options. Thus, offered a choice between a ticket guaranteeing $100 if a coin of unknown properties turns up "heads" and a corresponding ticket for "tails," most reasonable people would have no basis for choice—implying that 50 percent probability is attached to "tails" and 50 percent to "heads," although the coin is not known to be "fair."

A point that sometimes bothers students is that the probability distribution for the unknown parameter expresses beliefs, but says nothing explicitly about the strength with which the beliefs are held. This is a mistake, however—the strength or confidence of beliefs that the parameter will take on particular values is precisely expressed by the probability distribution. In the example above, the individual who placed 100 percent probability on the parameter P's being within the interval from .001 to .010 obviously has stronger or more confident beliefs about P than the individual who could attach only 50 percent probability to P's falling within this same range.

[8] This formula is strictly appropriate only for a discrete number of possible states of the world (values of the unknown parameter P). The following verbal statement of the formula may be helpful. The expected loss for any specified decision rule, EL (d.r.), is equal to the sum of a number of

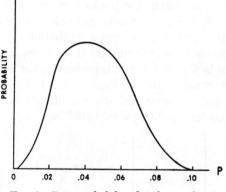

Fig. 3   Prior probability distribution for *P*.

A point of considerable importance here is the assumption that expected values of loss are sufficient to guide decision. If loss is measured in dollars, this implies, for example, that the individual in whose interest the analysis is conducted is indifferent between $500 certain, a 50 percent chance of $1000, a 5 percent chance of $10,000, or a 0.5 percent chance of $100,000. But we would not ordinarily regard, say, a small businessman as unreasonable if he took a loss of $600 certain, or even perhaps $1000 certain in the form of an insurance premium on property worth $100,000 where the contingency insured against had a known probability of 0.5 percent. On the other hand, it would seem unreasonable for General Motors to pay much more than the expected value as insurance on the (for it) very moderate loss contingency of $100,000. This much argument, if accepted, implies that where only "moderate" contingencies of loss are involved, expected values of dollar loss may be at least a roughly satisfactory guide.[9]

---

terms—one for each possible value, *P*, of the parameter—where each such term is the product of (1) the probability of error, given that *P* is the true parameter value; (2) the loss due to error, again given that *P* is the true value; and (3) the prior probability attached to *P* being the true value.

[9] A more theoretically satisfactory solution involves the substitution of a utility dimension for dollars in measuring "payoff" or loss. It has recently been demonstrated that, if certain very reasonable postulates are accepted,

Geometrically, the expected loss of a given decision rule could be pictured as the area under a curve showing, for each value of P on the horizontal axis, the product of the vertical heights for that value of P in Fig. 1 (conditional risk of error), Fig. 2 (conditional loss), and Fig. 3 (prior probability). However, the relationships are easier to interpret if we show, against the prior probabilities in Fig. 3, a Fig. 4 representing for each P the product of the vertical heights in Figs. 1 and 2. This product of the conditional risk of error and the conditional loss due to error will be called the conditional expected loss.

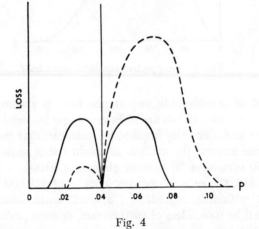

Fig. 4

Figure 4 reveals a rather important point: If the true P is near P_b (here $P_b = .04$), it does not matter much if we make the wrong decision (the conditional expected loss is small, even though the conditional probability of error is large, because the conditional loss near P_b in Fig. 2 is almost zero). This is to be contrasted with the strong emphasis that the classical approach places upon the risk of error α at the borderline or limiting value of the composite H_o.[10]

---

it is rational for individuals to calculate solely in terms of expected values of utility. More precisely, it is rational for the individual to act as if (1) he attaches a number, called a utility, to each possible (dollar) outcome; and (2) in choosing between probabilistic alternatives, he selects that one for which the expected value of utility is the higher.

[10] The Bayesian "break-even" value P_b would, most likely, be the limiting value of H_o for a classical analyst, although perhaps there might be

It may be useful to readers to work out a numerical solution for the example described above. Assuming a sample size of 50 with simple random sampling, the classical approach directs the analyst to choose his decision rule on the basis of only the information contained in Table 1—and not all of that, if he is supposed to concentrate his attention upon the α at $P = .04$ and the β at some *one* of the values within H_A, where we have provided two values ($P = .05$ and .06) to choose from. We may also remark that, if a conventional .05 level of significance is used, the conditional risk of Type II error will be in the neighborhood of 80 percent for $P = .06$ (90 percent for $P = .05$); if a .01 level of significance is used, the conditional risk of Type II error for each of these two alternatives is over 90 percent.

While the classical analysis requires a direct intuitive fixing of the determinants of the decision rule, the Bayesian approach builds up a simple and elegant structure for its determination. The information required is summarized in Table 2. The main

TABLE 2

BAYESIAN COMPUTATION TO FIND BEST DECISION RULE ρ_r, WHERE $n = 50$

	Conditional Probabilities of Error									Ex-pected Loss
	Type 1					Type II				
P	0	.01	.02	.03	.04	.05	.06	.07	.08	(EL)
$p_r = 0$	1.0	1.0	1.0	1.0	1.0	0	0	0	0	2.0000
.02	0	.3950	.6358	.7819	.8701	.0769	.0453	.0266	.0155	.6826
.04	0	.0894	.2642	.4447	.5995	.2794	.1900	.1265	.0827	.3852°
.06	0	.0138	.0784	.1892	.3233	.5405	.4162	.3108	.2260	.3984
.08	0	.0016	.0178	.0628	.1391	.7604	.6473	.5327	.4253	.5561
.10	0	.0001	.0032	.0168	.0490	.8964	.8206	.7290	.6290	.7288
Conditional loss	8	6	4	2	0	1	2	3	4	
Prior probability	0.1	.1	.1	.1	.2	.1	.1	.1	.1	

° Minimum EL.

some question on this point. The reason for identifying the two is that, to use our example, the classical approach speaks of a Type I "error" being committed if H_o is rejected when $P \le .04$, implying that H_o is the "correct" hypothesis or action in such cases, and that H_A is correct for $P > .04$. if the "correct" action is interpreted in a common-sense way as the choice involving smaller loss, the classical division between the composite H_o and H_A corresponds to the Bayesian division between that range for P for which one action is preferable (has less expected loss) and that for which the other is preferable. This makes P_b equivalent to the limiting value of H_o.

body of the table shows, for each decision rule considered, the conditional probability of error for values of the unknown parameter by hundreths from 0 to .08—of Type I error for $P \leqq .04$ and of Type II error for $P > .04$.[11] This part of the table corresponds to Fig. 1, except that to make the computations easy we shall allow only the nine discrete possible values for P shown. This limitation may be interpreted as an approximation; or, alternatively, it may simply be the case that only these discrete values are possible.

On the line below the main body of the table, the conditional losses of the two types of error are shown as a function of P. Since the conditional losses are independent of the decision rule, one line suffices to show them. This line corresponds to Fig. 2. The actual numbers are derived from the following expressions, where $L(R, P)$ is the conditional loss of rejecting as a function of P and $L(A, P)$ is the conditional loss of accepting as a function of P:

$$(\text{Type I})$$
$$L(R, P) = \begin{cases} 0, \text{ for } P \geqq .04 \\ 200 \, (.04 - P), \text{ for } P < .04 \end{cases}$$

$$(\text{Type II})$$
$$L(A, P) = \begin{cases} 0, \text{ for } P \leqq .04 \\ 100 \, (P - .04), \text{ for } P > .04 \end{cases}$$

The bottom line shows the prior probabilities $Pr_o\ (P)$. This line corresponds to Fig. 3; however, for simplicity of computation, the probabilities in the table are assumed to be uniform over the discrete values of P from 0 to .08, except for a bulge at $P = .04$.

Finally, the right-hand column of Table 2 shows the expected loss EL for each decision rule considered. The best decision rule is that for which EL attains its minimum: .3852 for the rule $p_r = .04$. As can be seen in Table 2 (or Table 1), this is equivalent to selecting an a of .5995, and a β of .2794 measuring at $P = .05$ or of .1900 measuring at $P = .06$, values for a and β that unaided

[11] It is immaterial how we treat the specific value $P = .04$ (i.e., whether we consider it to be part of H_o or of H_A). Since .04 is the break-even value P_b, the conditional loss of either decision under it is zero, so that whichever risk of error we consider will be canceled out when we multiply by the loss.

intuition would hardly be likely to hit upon using the classical approach.

Using this decision rule, we see that the sample result $p = .04$, for example, would lead to rejection of the null hypothesis, corresponding to rejection of the lot. This may seem surprising, since a population $P = .04$ would represent a (borderline) satisfactory lot. The explanation for the decision is that while, in this case, the prior probability distribution of Table 2 is symmetrical about $P = .04$, the loss function in Table 2 (see also Fig. 2) is not. Type I losses rise more rapidly than Type II losses, thus making us more willing to commit Type II errors than comparable Type I errors—speaking loosely, more inclined to reject the null hypothesis than to accept it.

Throughout this analysis we have assumed the sampling method and sample size n fixed; so, choice of a decision rule amounted to a choice of p_r. We shall only comment briefly on the consequences of changing n. It is immediately clear that n affects only the conditional risks of error (Fig. 1) which enter into both the classical and Bayesian procedures—Figs. 2 and 3 are unaffected. The selection of the sample size under the Bayesian approach follows directly from the basic principle of minimizing expected loss. In choosing the best decision rule for a given n, the minimum expected loss for that n, which we may denote EL^*, was determined. In principle, there is no difference finding EL^* for any n. It is necessary to establish whether an increase in n is justified, which will depend upon whether the reduction in expected loss achieved by the change in n is in excess of the additional sampling cost. The same principle applies to the choice of sampling method.

Probability: A Subjectivist View

L. J. SAVAGE

L. J. Savage is Professor of Statistics at Yale University.
This piece comes from his well-known book, *The Foundations of Statistics*, published in 1954.

1. INTRODUCTION

It is my tentative view that the concept of personal probability is, except possibly for slight modifications, the only probability concept essential to science and other activities that call upon probability. I propose to discuss the shortcomings I see in that particular personalistic view of probability, which, for brevity, shall here be called simply "the personalistic view"; to point out briefly the relationships between it and other views; to criticize other views in the light of it; and to discuss the criticisms holders of other views have raised, or may be expected to raise, against it.

2. SOME SHORTCOMINGS OF THE PERSONALISTIC VIEW

I can answer, to my own satisfaction, some criticisms of the personalistic view that have been brought to my attention. These points are discussed later in the chapter, but in this section I state and discuss as clearly as I can those that I find more difficult and confusing to answer.

According to the personalistic view, the role of the mathematical theory of probability is to enable the person using it to detect inconsistencies in his own real or envisaged behavior. It is also understood that, having detected an inconsistency, he will remove it. An inconsistency is typically removable in many different ways, among which the theory gives no guidance for choosing. Silence on this point does not seem altogether appropriate, so there may be room to improve the theory here. Consider an example: The person finds on interrogating himself about the possible outcome of tossing a particular coin five times that he considers each of the 32 possibilities equally probable, so each has for him the numerical probability 1/32. He also finds that he considers it more probable that there will be four or five heads in the five tosses than that the first two tosses will both be heads. Now, reference to the mathematical theory of probability soon shows the person that, if the probability of each of the 32 possibilities is 1/32, then the probability of four or five heads out of five is 6/32, and the probability that the first two tosses will be heads is 8/32, so the person has caught himself in an inconsistency. The theory does not tell him how to resolve the inconsistency; there are literally an infinite number of possibilities among which he must choose.

In this particular example, the choice that first comes to my mind and I imagine to yours, is to hold fast to the position that all 32 possibilities are equally likely and to accept the implications of that position, including the implication that four or five heads out of five is less probable than two heads out of two. I do not think that there is any justification for that choice implicit in the example as formally stated, but rather that in the sort of actual situation of which the example is a crude schematization there generally are considerations not incorporated in the example that do justify, or at any rate elicit, the choice.

To approach the matter in a somewhat different way, there seem to be some probability relations about which we feel relatively "sure" as compared with others. When our opinions, as reflected in real or envisaged action, are inconsistent, we sacrifice the unsure opinions to the sure ones. The notion of "sure" and "unsure" introduced here is vague, and my complaint is precisely that neither the theory of personal probability, as it is developed in this book, nor any other device known to me renders the notion less vague. There is some temptation to introduce probabilities of a second order so that the person would find himself saying such things as "the probability that B is more probable than C is greater than the probability that F is more probable than G." But such a program seems to meet insurmountable difficulties.

It may be clarifying, especially for some readers under the sway of the objectivistic tradition, to mention that, if a person is "sure" that the probability of heads on the first toss of a certain penny is one-half, it does not at all follow that he considers the coin fair. He might, to take an extreme example, be convinced that the penny is a trick one that always falls heads or always falls tails.

Logic, to which the theory of personal probability can be closely paralleled, is similarly incomplete. Thus, if my beliefs are inconsistent with each other, logic insists that I amend them, without telling me how to do so. This is not a derogatory criticism of logic but simply a part of the truism that logic alone is not a complete guide to life. Since the theory of personal probability is more complete than logic in some respects, it may be somewhat disappointing to find that it represents no improvement in the particular direction now in question.

A second difficulty, perhaps closely associated with the first one, stems from the vagueness associated with judgments of the magnitude of personal probability. The postulates of personal probability imply that I can determine, to any degree of accuracy whatsoever, the probability (for me) that the next president will be a Democrat. Now, it is manifest that I cannot really determine that number with great accuracy, but only roughly. Since, as is widely recognized, all the interesting and useful theories of modern science, for example, geometry, relativity, quantum mechanics, Mendelism, and the theory of perfect competition, are inexact; it may not at first sight seem disquieting that the theory of personal probability should also be somewhat inexact. As will immediately be explained, however, the theory of personal probability cannot safely be compared with ordinary scientific theories in this respect.

I am not familiar with any serious analysis of the notion that a theory is only slightly inexact or is almost true, though philosophers of science have perhaps presented some. Even if valid analyses of the notion have been made, or are made in the future, for the ordinary theories of science, it is not to be expected that those analyses will be immediately applicable to the theory of personal probability, normatively interpreted; because that theory is a code of consistency for the person applying it, not a system of predictions about the world around him.

3. CRITICISM OF OTHER VIEWS

It will throw some light on the personalistic view to say briefly how some other views seem to compare unfavorably with it.

It is one of my fundamental tenets that any satisfactory account of probability must deal with the problem of action in the face of uncertainty. Indeed, almost everyone who seriously considers probability, especially if he has practical experience with statistics, does sooner or later deal with that problem, though often only tacitly. Even some personalistic views seem to me too remote from the problem of action, or decision.

Keynes, writing in 1921 of what are here called objectivistic views, complained, "The absence of a recent exposition of the logical basis of the frequency theory by any of its adherents has been a great disadvantage to me in criticizing it." I believe that his complaint applies as aptly to my position today as to his then, though I cannot pretend to have combed the intervening literature with anything like the thoroughness Keynes himself would have employed. Reichenbach, to be sure, presents in great detail an interesting view that must be classified as objectivistic, but it seems far removed from those that dominate modern statistical theory and form the main subject of the following discussion. Whatever objectivistic views may be, they seem, to holders of necessary and personalistic views alike, subject to two major lines of criticism. In the first place, objectivistic views typically attach probability only to very special events. Thus, on no ordinary objectivistic view would it be meaningful, let alone true, to say that on the basis of the available evidence it is very improbable, though not impossible, that France will become a monarchy within the next decade. Many who hold objectivistic views admit that such everyday statements may have a meaning, but they insist, depending on the extremity of their positions, that that meaning is not relevant to mathematical concepts of probability or even to science generally. The personalistic view claims, however, to analyze such statements in terms of mathematical probability, and it considers them important in science and other human activities.

Secondly, objectivistic views are, and I think fairly, charged with circularity. They are generally predicated on the existence in nature of processes that may, to a sufficient degree of approximation, be represented by a purely mathematical object, namely, an infinite sequence of independent events. This idealization is said, by the objectivists who rely on it, to be analogous to the treatment of the vague and extended mark of a carpenter's pencil as a geometrical point, which is so fruitful in certain contexts. When it is pointed out to the objectivist that he uses the very theory of probability in determining the quality of the approximation to which he refers, he retorts that the applied geometer —a fictitious character whose reputation for solidity in science is unquestioned—likewise uses geometry in determining the quality of his approximations. Let the geometer then be challenged, and he replies with a threefold reference to experience, saying, "It is a common experience that with sufficient experience one develops good judgment in the use of geometry and thenceforth generally experiences success in the predictions he bases on it." "Now," says the objectivist, "the geometer's answer is my answer." But it seems to critics of objectivistic views that, though the geometer may be entitled to make as many allusions to experience as he pleases, the probabilist is not free to do so, precisely because it is the business of the probabilist to analyze the concept of experience. He, therefore, cannot properly support his position by alluding to experience until he has analyzed that concept, though he can, of course, allude to as many experiences as he wishes.

4. THE ROLE OF SYMMETRY IN PROBABILITY

An important and highly controversial question in the foundations of probability is whether and, if so, how symmetry considerations can determine the probabilities of at least some events.

Symmetry considerations have always been important in the study of probability. Indeed, early work in probability was dominated by the notion of symmetry, for it was usually either concerned with, or directly inspired by, symmetrical gambling apparatus such as dice or cards. To illustrate those classical problems, suppose that a gambler is offered several bets concerning the possible outcome of rolling three dice, where it is to be understood that refraining from any bets at all may be among the available "bets." Which of the available bets should the gambler choose? Perhaps I distort history somewhat in insisting that early problems were framed in terms of choice among bets, for many, if not most, of them were framed in terms of equity, that is, they asked which of two players, if either, would have the advantage in a hypothetical bet. But, especially from the point of view of the earlier probabilists, such a question of equity is tantamount to a question of choice among bets, for to ask which of two "equal" betters has the advantage is to ask which of them has the preferable alternative, as was pointed out quite explicitly by D. Bernoulli.

In effect, the classical workers recommended the following solution to the problem of three dice, with corresponding solutions to other gambling problems:

1. Attach equal mathematical probabilities to each of the 216 ($= 6^3$) possible outcomes of rolling the three dice. (There are 6^3 possibilities, because the first, second, and third dice can each show any of six scores, all combinations being possible.)
2. Under the mathematical probability established in Step 1, compute the expected winnings (possibly negative) of the gambler for each available bet.
3. Choose a bet that has the largest expected winnings among those available.

At present it is appropriate to refrain from criticisms of the use made of expected winnings until the next chapter and to concentrate discussion on the notion that the 216 possibilities should be considered equally probable, which can conveniently be done by drastically reducing the class of bets considered to be available. Say, for definiteness, that the only bets to be considered are simply even-money bets of one dollar, that the triple of scores falls in a preassigned subset of the 216 possibilities. When attention is focused on this restricted class of bets, the total recommendation is seen to imply that the probability measure defined in the first step of the recommendation be adopted as the personal probability of the gambler. To put it differently, a gambler who adopts the recommendation will hold the 216 possible outcomes equally probable not only in some abstract sense, but also in the sense of personal probability as defined here.

The notion that the 216 possibilities should be regarded as equally probable is familiar to everyone; for it is taken for granted wherever gentlemen gamble as well as in the standard high-school algebra courses, where it serves to illustrate the theory of combinations and permutations.

Traditionally, the equality of the probabilities was supposed to be established by what was called the principle of insufficient reason, thus: Suppose that there is an argument leading to the conclusion that one of the possible combinations of ordered scores, say [1, 2, 3], is more probable than some other, say [6, 3, 4]. Then the information on which that hypothetical argument is based has such symmetry as to permit a completely parallel, and therefore equally valid, argument leading to the conclusion that [6, 3, 4] is more probable than [1, 2, 3]. Therefore, it was asserted, the probabilities of all combinations must be equal.

The principle of insufficient reason has been and, I think, will continue to be a most fertile idea in the theory of probability; but it is not so simple as it may appear at first sight, and criticism has frequently and justly been brought against it. Holders of necessary views typically attempt to put the principle on a rigorous basis by modifying it in such a way as to take account of such criticism. Holders of personalistic and objectivistic views typically regard the criticism as not altogether refutable, so they do not attempt to establish a formal postulate corresponding to the principle but content themselves—as I shall here—with exhibiting an element of truth in it.

One of the first criticisms is that the principle is not strictly applicable for a person who has had any experience with the apparatus in question, or even with similar apparatus. Thus, attempts to use the principle, as I have stated it, to prove that there is no such thing as a run of luck at dice, as actually played, are invalid. The person may have had relevant experience, directly or vicariously, not only with gambling apparatus itself, but also with people who make and handle it, including cheaters.

It is not always obvious what the symmetry of the information is in a situation in which one wishes to invoke the principle of insufficient reason. For example, d'Alembert, an otherwise great 18th-century mathematician, is supposed to have argued seriously that the probability of obtaining at least one head in two tosses of a fair coin is 2/3 rather than 3/4. "Heads," as he said, might appear on the first toss, or, failing that, it might appear on the second, or, finally, might not appear on either. D'Alembert considered the three possibilities equally likely.

It seems reasonable to suppose that, if the principle of insufficient reason were formulated and applied with sufficient care, the conclusion of d'Alembert would appear simply as a mistake. There are, however, more serious examples. Suppose, to take a famous one, that it is known of an urn only that it contains either two white balls, two black balls, or a white ball and a black ball. The principle of insufficient reason has been invoked to conclude that the three possibilities are equally probable, so that in particular the probability of one white and one black ball is concluded to be 1/3. But the principle has also been applied to conclude that there are four equally probable possibilities, namely, that the first ball is white and the second also, that the first is white and the second black, etc. On that basis, the probability of one white and one black ball is, of course, 1/2. Personally, I do not try to arbitrate between the two conclusions but consider that the existence of the pair of them reflects doubt on the notion that a person's knowledge relevant to any matter admits any full and precise description in terms of propositions he knows to be true and others about which he knows nothing.

Most holders of personalistic views do not find the principle of insufficient reason compelling, because they envisage the possibility that a person may consider one event more probable than another without having any compelling argument for his attitude. Viewed practically, this position is closely associated with the first criticism of the principle of insufficient reason, for the holder of a personalistic view typically supposes that the person is under the influence of experience, and possibly even biologically determined inheritance, that expresses itself in his opinions, though not necessarily through compelling argument.

Holders of personalistic views do see some truth in the principle of insufficient reason, because they recognize that there are frequently partitions of the world, associated with symmetrical-looking gambling apparatus and the like, that many and diverse people all consider (very nearly) uniform partitions. As was illustrated in the preceding section, we often feel more "sure" about probabilities derived from the judgment that such partitions are uniform than we do about others. Such partitions are, moreover, very important in that they provide some events the probability of which to diverse people is in agreement. Though the events concerned are often of no importance in themselves, agreement about them can, through the statistical invention of randomization, contribute to agreement about all sorts of issues open to empirical investigation. Widespread though the agreement about the near uniformity of some partitions is, holders of personalistic views typically do not find the contexts in which such agreement obtains sufficiently definable to admit of expression in a postulate.

Holders of purely objectivistic views see no sense at all in the original formulation of the principle of insufficient reason, for it uses "probability" in a manner they consider meaningless. But they too see an element of truth in the principle, which they consider to be established as a part of empirical physics. Thus, for example, they regard it as an experimental fact, admitting some explanation in terms of theoretical physics, that three dice manufactured with reasonable symmetry will exhibit each of the 216 possible patterns with nearly equal frequency, if repeatedly rolled with sufficient violence on a suitable surface.

Holders of personalistic views agree that experiments or, more generally, experiences determine to a large extent when people employ the idea of insufficient reason. Thus, though experiments with gambling apparatus, quite apart from gambling itself, have a fascination that perhaps exceeds their real interest, such experiments are not altogether worthless. On the one hand, they provide strong evidence that a person cannot expect to maintain a symmetrical attitude toward any piece of apparatus with which he has had long experience, unless he is virtually convinced at the outset that the possible states of the apparatus are equally probable and independent from trial to trial. To say it in the more familiar and sometimes more congenial language of objective probability, long experiments with coins, dice, cards, and the like have always shown some bias, and often some dependence from trial to trial. On the other hand (and this has the utmost practical importance), it has been shown that, with skill and experience, gambling apparatus, or its statistical equivalent, can be manufactured in which the bias and the dependence from trial to trial are extremely small. This implies that groups of very diverse people can be brought to agree that repeated trials with certain apparatus are nearly uniform and nearly independent. Thus certain methods of obtaining random numbers and other outcomes of uniform and independent trials, which are vital to many sorts of experimentation, have justifiably found acceptance with the scientific public.

SAMPLING AND APPLICATIONS OF t, x^2, AND F TESTS

Business and government are continually engaged in activities where sampling can be used to reduce the cost of obtaining information. For example, in production management, sampling is often used to test and maintain the quality of materials and final product. In the first article, Morris Hansen and William Hurwitz discuss the advantages and disadvantages of using probability samples rather than "quota" or other judgment samples in survey sampling. They describe the way in which probability samples are designed, the cost of probability samples, the use of probability samples in prediction, and the choice of sampling methods. Their paper is based on many years of experience as principal statisticians at the United States Bureau of the Census.

Sampling techniques are applicable, of course, to other than human populations. In the next article, John Neter describes a number of applications of statistical techniques in the area of accounting, particular attention being given to auditing. He describes the use of statistical techniques to control clerical accuracy in the Census Bureau, as well as other similar control chart techniques used by United Air Lines and other companies. Then he describes the application by firms of statistical sampling techniques to accounting records and to physical property.

Three of the most fundamental statistical tests are the t, χ^2, and F tests. The purpose of the next two articles in this part is to describe briefly some applications of these tests, which are used repeatedly in economics and business. The t test, when first considered in most elementary courses, is used to determine whether a mean equals a specified figure.[1] The article by Lawrence Fouraker and Stanley Siegel uses this test to see whether, in a situation of bilateral monopoly (a monopolistic seller dealing with a monopsonistic buyer), contracts tend to be negotiated so that joint profits are maximized. In addition, the same kind of test is used to see whether two other hypotheses—the marginal intersection hypothesis and the Fellner hypothesis—hold in a situation of this sort. Fellner's hypothesis is that the equilibrium price will depend on the relative bargaining strength of the buyer and seller. The marginal intersection hypothesis is that equilibrium will be achieved at the intersection of the marginal functions in Fig. 2 of the paper. The results are interesting both as an illustration of the use of the test and as an example of the use of laboratory experiments in economics.

The χ^2 test is often used to determine whether there is independence in a contingency table. For example, suppose that firms are classified both by their size at the beginning of a period and by their percentage growth during the period. Then one can use the χ^2 test to see whether the distribution of growth rates is the same regardless of initial size. In the next article, the editor carries out this test for the steel, petroleum, and rubber tire industries, the purpose being to see whether Gibrat's law, which has often been used in models of industry structure and behavior, holds. According to Gibrat's law, the probability of a given proportionate change in size during a specified period is the same for all firms in a given industry—regardless of their size at the beginning of the period. Also, the F test, which tests whether or not two variances are equal, is applied to determine whether the variance of the growth rates of the small firms is equal to the variance of the growth rates of the large firms.[2]

---

[1] Of course, tests based on the t distribution are used in a great many other areas as well. For example, the article by Moore illustrates how it is used in simple regression.

[2] Of course, tests based on the F distribution are used to test hypotheses other than the equality of variances.

Dependable Samples For Market Surveys

MORRIS H. HANSEN and WILLIAM N. HURWITZ

Morris Hansen and the late William Hurwitz were prominent statisticians in the United States Bureau of the Census. Their paper appeared in the *Journal of Marketing* in 1949.

There has been considerable discussion, in the market research field, of the advantages or disadvantages of adopting probability samples instead of the "quota" or other judgment sampling methods that have been widely used. Apparently both approaches are now being used extensively in commercial work.

It is reasonable to assume that quota and other judgment methods are used in many instances where in fact an appropriately designed probability method would give results of greater reliability at equal cost. Perhaps more important, there are many instances in which a probability sample would, if the facts were known, cost more for achieving results of equal reliability, but where the use of a probability sample is desirable simply because the probability sampling method, when properly carried through, gives results of known sampling precision, whereas the sampling precision of the results of a quota or other judgment sample cannot be established objectively but depends upon various assumptions and judgments that are more or less difficult to defend. It is no doubt true, at the other extreme, that probability samples are used in many situations where their use involves additional expense that is not justifiable and where quota or other judgment sampling methods would have served satisfactorily at lower cost.

It is to be emphasized that, while this paper describes the feasibility and desirability of using probability samples, it is not intended to imply that only probability samples should be used. There is need for careful review and consideration of whether a probability sample is best in a particular circumstance, or whether the additional insurance as to reliability provided by a probability sample is worth what it may cost. As a general rule, it should be thought of as worth while to take a probability sample where results of high precision are needed, or where objective and unbiased results are wanted because important decisions or courses of action will be determined on the basis of the sample results. The investigator should realize that only the sampling error is controlled by the use of a probability sample. There are other sources of error in surveys, and these may be more important in many instances than the sampling error. At the same time, when considerable resources are invested in a survey, and when careful survey procedures are laid out and followed, the recognition that sources of error other than those arising from sampling are present in a survey does not justify the use of loose sampling methods.

With a probability sample, properly designed and executed, one can, by taking a large enough sample, achieve results from a sample that will be as close as desired to the results obtainable from a complete census taken under the same conditions. With a sample of moderate size, one can achieve results whose precision, in terms of range of error around the results of a complete census, can be established with confidence. The magnitude of the sampling errors is determined by the design and size of the sample.

This paper discusses briefly how the costs of probability sampling arise, their general magnitude, and the importance of paying the necessary costs for obtaining results of measurable sampling reliability where results of high precision are needed, and also illustrate how probability methods can be applied in practice. There is no attempt to distinguish here those situations in which it is worth "buying the insurance" of measurable sampling errors in survey results through the use of probability samples. We regard this as an important subject that needs fuller consideration than can be given here.

HOW TO PLAN A PROBABILITY SAMPLE

The rules for getting a probability sample require neither a mathematical formula nor complex procedures. For example, to obtain a probability sample of 2 percent of the blocks in a city, one could number serially the blocks of a city map, and draw a random number between 1 and 50. Assume this random number is 7. Then if the 7th, 57th, 107th, etc., blocks are included in the sample, and a census is taken of the population residing in these sample blocks, the result would be a probability sample of the people resident in the city. A variation in procedure, still simple, would be to take, say, a 10 percent sample of blocks drawn in the manner described above, make a complete listing of the households in the selected blocks, and include in the sample every fifth household from this listing. Again, the result would be a 2 percent probability sample of people. These are illustrations of probability samples. A cursory examination of the simple steps described above to obtain a probability sample might give one cause for wondering why such a simple procedure should cost more per interview than the quota or other types of judgment sampling commonly used.

Cost of Probability Samples

Perhaps a consideration of what probability sampling calls for in the way of extra work or inconvenience that is not always called for in the other methods may indicate why one would expect the cost to be higher per interviewer, that is, higher for a given *size* of sample.

With a probability sample, the enumerator may have to make a number of calls in order to complete an interview. He will have to go to predesignated blocks and to predesignated households (1) to obtain an interview, and is not permitted the discretion of substituting a more accessible household when no one is found at home on first call. He may have to climb stairs and walk through back alleys and go over poor roads in rural areas in order to meet the requirements of the probability sample. The rules for obtaining a probability sample, though they may appear to be arbitrary to the enumerator, must be adhered to closely if one wants to be sure that an unbiased cross section of the population is covered, and wants to be able to measure the amount of sampling variability in the results.

(2) Another cause for the additional cost in the probability sampling illustrated above is the need for designating the sample blocks and for listing all of the households in these blocks. From these lists the sample required is drawn.

In a quota or judgment sample one faces the risk of not obtaining the appropriate representation of the persons not at home on first call, or of persons living in the relatively harder-to-get-at places, or of any class that is inconvenient or which in the judgment of the enumerator should not be included in the sample. One pays added costs in a probability sample to get the proper representation of classes of the population for which it is impractical to set separate quotas or to depend on the judgment of the enumerator to obtain the proper representation.

It is important to note that, with probability samples, it is possible to specify in advance a design that meets the accuracy requirements. Thus, the number of households required can be specified reasonably well in advance for any particular probability sample. On the other hand, there is no certainty with quota or judgment sampling that an increase in sample size will yield an increase in accuracy. To be able to state fairly accurately the reliability of the sample results compared to the results of a census of the population, without actually taking a census to make this comparison, may be worth a considerable added cost over a procedure which can only be validated by a complete census.

Now let us examine whether the costs of taking a probability sample are beyond the means of people in the marketing field. First, let us consider some of the main aspects of the additional costs that may be involved in a probability sample if the particular sampling methods described above were followed.

1. One cost is the objective designation of the sample. In the sample design described above, this includes numbering the blocks on a map and listing the households on the selected

blocks, and selecting the sample households from this list.

2. Next is the cost of interviewing and of following up households to the point where interviews are obtained from substantially all of the designated households. In the Census Bureau it is usually assumed that, if the required information is obtained from more than 95 percent of the designated households, one is entitled to feel fairly secure in assuming that the sample was taken in conformance with sampling theory, even though assumptions may be necessary for the remaining 5 percent. It has been found that for some purposes trouble arises even when making assumptions for only 5 percent.

3. A third cost is involved in careful supervision and checking to insure that the specified steps are carried out substantially as specified.

There are numerous illustrations in the work of the Census Bureau of the cost of these procedures. As one example, in about 40 surveys of population and dwelling unit characteristics for individual cities taken during 1947, the average cost per household was approximately $2, including both field and office costs. This was for a survey in which the interview could be with a responsible member of the household rather than a specified individual. Each city involved about 3500 interviews. In these surveys, the schedule was a relatively simple one. In other surveys, where the schedule is more complex, or if a more complex sampling procedure is used, the average cost per household may run from $3 to $6, or considerably higher. Note that in these higher costs surveys involving long interviews the additional cost required in using a probability sample rather than a judgment sample is less, since the costs of selecting the sample and of calling back becomes smaller in relation to the cost of interviewing.

The previous discussion gives some basis for evaluation of the range of the costs per interview that may be expected in jobs based on probability samples. It should be emphasized, however, that it is not cost *per interview,* but costs for a result of a given reliability that counts. Moreover, if important purposes are to be served by the survey, and reliable data are needed on which to base decisions, it is worthwhile paying considerably more, if necessary, for data of known sampling reliability.

LIMITATIONS OF QUOTA SAMPLES

Methods that are subject to personal bias, and whose reliability cannot be objectively evaluated sometimes lead to considerable difficulty. The failure of the election polls may be a significant illustration. In that instance, there is reason to believe that the use of judgment-sampling methods that were subject to more or less serious biases was at least one of the important causes of the difficulty. Another illustration may be cited from the experience of the Bureau of the Census, indicating how a judgment sampling method that was used earlier in the census gave misleading results that were finally corrected with a probability sample.

During the early stage of the war, the Bureau was using a carefully controlled sampling method involving fixed quotas in the survey from which monthly information on the total size of the labor force, employment and unemployment, hours worked, and other characteristics of the population were reported. The method originally used in this survey was objective to the extent that a predesignated area sample was used in which the enumerator had no choice in determination of who was to be included. Nevertheless, judgment was involved in the final selection made, in that in predesignating the areas and dwelling units to be included in the sample, quotas were set on the number of interviews that would be taken as between rural and urban areas. In addition, certain rules that presumably insured proportionate representation but that violated probability sampling principles were introduced for selecting the particular households to be included from the sample of areas. Thus, the method did not insure a fixed probability of including each household in the sample. The results of this sample showed farm employment figures during the early period of the war at about a constant level for a period of a year or two until probability sampling methods of the sort described above were introduced. The probability sample revealed that a very substantial and significant decline had taken place in agricultural employment during the period.

This is an illustration of how a method that appeared to be

sound at the time it was used, but that lacks the precision of probability sampling, may sometimes give seriously misleading results. Only methods that can be strongly defended could have withstood the attacks and criticisms in such a period. Actually, the major declines in farm employment and farm population shown by the probability sample were subsequently confirmed by the 1945 Census of Agriculture.

USE OF PROBABILITY SAMPLES IN PREDICTION

The Bureau had a related but quite different experience immediately after the close of the war when the unemployment estimates were under serious attack. Many people had predicted there would be about eight million unemployed during this reconversion period when war contracts were canceled, but the survey showed less than two million. Had it not been that the reliability and the unbiased character of the sample results could be defended with complete confidence, and by objective evidence, there would have been much less assurance in the published figures and no doubt it would have taken a much longer time to accomplish an adjustment of national policy to conform with the real facts.

The Bureau often faces the problem of making predictions from a sample of results which subsequently become available from a census. One of the important uses of sampling in the census is to draw a sample from the census returns and to publish estimates of what the census will show long in advance of the time the complete census results can be compiled. When such estimates are made from a probability sample, it is possible to compute the reliability of the sample estimates, based on probability theory, and to publish measures of reliability with the publication of the estimates themselves. In this way, we are again and again up against the test of being able to make theory and practice conform, and of having sample estimates meet the acid test of subsequent complete census publication of estimated figures. The subsequent comparisons of actual sampling errors with those predicted by the mathematical theory show that the specified procedures are followed closely enough to make mathematical theory applicable. As an example, the 1945

Census of Agriculture provides an illustration of how it is possible to design samples that will give results of predictable reliability. On the basis of tabulations from the sample of the returns, the Bureau in July 1946 published national estimates for 61 agricultural items, together with a statement of the precision of each estimate. Corresponding figures from the complete Census of Agriculture became available about a year later. The estimates and their sampling errors as originally published, together with the relative differences between the sample estimates and the complete census returns appears in Table 1. It is seen that the complete census was in reasonable agreement with the advance statements of the precision of the original estimates. Three (5 percent) of the 61 differences between sample estimate and census exceeded two standard deviations, and none exceeded three standard deviations.

It is important to note how these comparisons of sample estimates with subsequent complete census results differ from the supposed validation of sample election polls by citing the relative success of such polls in previous elections. The essential difference is that the predictions of the reliability of the results of these probability samples are founded on mathematical theory and on the use of survey methods where theory and practice are in substantial conformance, rather than merely on the fact that similar surveys have given good results in the past. In fact, the Agriculture Census sample estimates, for example, were made without having any prior experience involving the prediction from a sample of what a census of agriculture would show. It was known on the basis of the procedures followed in drawing the sample and preparing estimates from it, and the mathematical theory appropriate to these procedures, that the results would come out about as predicted. If the process were repeated, similar results would have been obtained.

THE CHOICE OF SAMPLING METHODS

It was pointed out earlier that efficient sampling is accomplished by adapting the methods used to the particular sampling problem. Listed below are the criteria that are followed in the application of probability sampling in the Bureau of

TABLE 1.
PRELIMINARY SAMPLE ESTIMATE OF FARMS, FARM CHARAC-
TERISTICS, AND VALUE OF FARM PRODUCTS FOR 1945, COEFFI-
CIENT OF VARIATION OF THE SAMPLE ESTIMATES, AND
PERCENT DEVIATIONS OF THE SAMPLE ESTIMATES FROM
FINAL RESULTS—1945 CENSUS OF AGRICULTURE

(The sample estimates and their estimated coefficient of variation were published in a
census release dated July 30, 1946. The corresponding complete census figures became
available about a year later.)

Item	Sample Estimate	Coeff. of Variation of Sample Estimate (Percent)	Deviation of Estimate from Complete Census Results (Percent)
Farms, number	5,877,000	0.5	0.3
Land in farms, acres	1,148,355,000	0.5	0.6
Cropland harvested, acres	343,396,000	2.0	−2.7
Farm operators—			
By residence:			
Residence on farm operated, number	5,469,000	1.0	0.2
Residence not on farm operated, number	341,000	4.0	1.2
By tenure:			
Full owners and managers, number	3,308,000	1.0	−1.0
Part owners, number	668,000	2.0	1.1
All tenants, number	1,901,00	1.0	2.3
By color and tenure:			
All white operators, number	5,179,00	0.5	0.2
Full owners and managers, number	3,132,000	1.0	−1.0
Part owners, number	639,000	2.0	1.5
All tenants, number	1,408,000	1.0	2.3
All nonwhite operators, number	698,00	3.0	1.3
Full owners and managers, number	176,000	4.0	0.1
Part owners, number	29,000	6.0	−5.7
All tenants, number	493,000	3.0	2.1
By age:			
Under 35 years, number	1,000,000	2.0	0.7
35 to 54 years, number	2,780,000	1.0	0.1
55 to 64 years, number	1,170,000	1.0	−0.2
65 years and over	849,000	2.0	−0.1
By work off farm:			
All operators reporting, number	1,569,000	2.0	−0.1

TABLE 1.—(cont.)

Item	Sample Estimate	Coeff. of Variation of Sample Estimate (Percent)	Deviation Estimate from Complete Census Results (Percent)
Operators reporting:			
1 to 49 days, number	313,000	3.0	0.1
50 to 99 days, number	190,000	3.0	6.5
100 to 149 days, number	126,000	3.0	1.6
150 to 199 days, number	122,000	3.0	1.3
200 to 249 days, number	151,000	4.0	0.0
250 days and over, number	667,000	4.0	−2.4
Specified facilities in farm dwelling:			
Electricity, farms reporting	2,835,000	3.0	1.7
Radio, farms reporting	4,237,000	2.0	−0.6
Telephone, farms reporting	1,868,000	2.0	0.1
Motortrucks on farms, farms reporting	1,274,000	2.0	−2.0
Number	1,460,000	2.0	−2.0
Tractors on farms, farms reporting	2,001,000	2.0	−0.1
Number	2,425,000	2.0	0.1
Farms by size:			
Under 10 acres, number	561,000	4.0	−5.6
10 to 29 acres, number	952,000	2.0	0.3
30 to 49 acres, number	722,000	1.0	1.9
50 to 69 acres, number	481,000	1.0	1.8
70 to 99 acres, number	691,000	1.0	0.9
100 to 139 acres, number	640,000	1.0	1.0
140 to 179 acres, number	568,000	1.0	0.4
180 to 219 acres, number	284,000	1.0	0.4
220 to 259 acres, number	212,000	1.0	0.8
260 to 499 acres, number	483,000	2.0	2.1
500 to 999 acres, number	173,000	3.0	−0.4
1,000 acres and over, number	110,000	4.0	−2.6
Farms by total value of farm products:			
Under $250, number	548,000	4.0	−0.8
$250 to $399, number	419,000	4.0	−3.4
$400 to $599, number	496,000	4.0	−3.5

TABLE 1.—(cont.)

Item	Sample Estimate	Coeff. of Variation of Sample Estimate (Percent)	Deviation Estimate from Complete Census Results (Percent)
$600 to $999, number	769,000	2.0	−1.5
$1,000 to $1,499, number	724,000	2.0	0.8
$1,500 to $2,499, number	928,000	2.0	2.1
$2,500 to $3,999, number	760,000	2.0	2.3
$4,000 to $5,999, number	520,000	3.0	1.2
$6,000 to $9,999, number	404,000	4.0	1.4
$10,000 and over, number	292,000	5.0	1.0
Farms producing products primarily for sale, number	4,469,000	1.0	0.1
Farms producing products primarily for own household use, number	1,301,000	4.0	0.9
Total value of farm products sold or used by farm households, dollars	18,345,567,000	3.0	1.3
Value of farm products sold, dollars	16,496,282,000	3.0	1.6
Value of farm products used by farm households, dollars	1,849.285,000	2.0	−1.5

the Census, and that provide a guide to the choice of effective methods. These criteria are as follows:

1. Use sampling methods for which one can get from the sample itself an objective measure of the precision of the sample estimates.

2. Use only simple, straightforward procedures, and insist on adequate field supervision and control to make sure that the work is carried out in substantial conformance with the specifications. When this is accomplished, close conformance may be obtained between theory and practice.

3. From among the alternative methods that meet the first two criteria, methods should be used that meet necessary time schedules and other administrative restrictions, and that yield results of maximum reliability per dollar of cost. Sampling theory provides powerful tools for accomplishing this. It does not provide a unique guide to the best sample but it does give effective guidance in arriving at comparatively efficient samples. At the

same time, it provides accurate measures of the precision of the results actually obtained.

In getting at the optimum sample design, it has been assumed that the job is to estimate from a sample what would have been obtained from a complete census. This statement of the problem avoids dealing with response errors; i.e., errors that are present in a census taken with equal care. In the practical implications of survey design, however, one should take into account interviewing and response errors also, and allocate resources among sampling, size and design of sample, and interviewing techniques, etc., that will take joint account of response errors and sampling errors and minimize the combined effect of these two. To the extent that one has the ability to measure and take steps to control response errors, this is a comparatively simple mathematical problem. The real problem in this regard is the measurement of response errors—how they arise and how they can be controlled. Nevertheless, decisions on sample survey design often involve assumptions as to the joint effect of response errors and sampling errors. Insofar as feasible the techniques chosen should be those that will succeed in the joint minimization of the two rather than deal only with one source of error or the other. There is urgent need for fuller study of the sources of non-sampling errors in survey results and of methods for their measurement and control.

As has already been suggested, in any practical sampling design problem there are many alternative ways a sample might be chosen. The problem is to explore all the resources and techniques available and choose from among these in accordance with the three criteria listed above. Thus, one might find that, at the same fixed cost he could get alternatively, a sample of 4000 households from a city by taking 200 blocks and using a subsampling ratio that will yield an average of 20 dwelling units per block; or a sample of 3600 households by taking a sample of 400 blocks with an average of 9 dwelling units per block; or a sample of 3000 households by taking a sample of 1,000 blocks with an average of 3 dwelling units per block, etc. Then the job is to pick the one from among these and other similar alternatives that will give the most reliable results for the fixed cost.

Sampling theory provides assistance in doing this. In many practical problems for estimating general family or personal characteristics of a population, it turns out that the optimum design involves taking somewhere between 2 and 8 households from a sample block on the average and sufficient blocks to achieve the necessary reliability. However, the appropriate specifications vary with the nature of the survey and the particular information that is being collected so that no rules can be given that will be applicable in all situations.

A variation in design that is something desirable is to use separate sampling ratios for large and small blocks.

There are quite different ways in which a probability sample from a city might be drawn. Thus, if there is a moderately up-to-date city directory containing a list of the addresses in the city, such a directory can be used effectively for drawing the sample. Moreover, it can be used in such a way that, even though the directory is not complete and up-to-date, one can get an unbiased sample of all dwelling units in the area, including those not listed in the directory. Thus we can think of the population of dwelling units in an area as divided into two classes. One class consists of those dwelling units that are listed in the directory, another class consists of those that are not. Then the sampling procedure might be to draw from the directory a sample of those listed in the directory by taking, say, every 50th dwelling unit in the directory, or perhaps by taking clusters of 3 and skipping 147, or by following whatever the formula needs to be in order to get an efficient sample for this particular job. This gives a sample for that part of the universe that is listed in the directory. For the remaining part of the universe consisting of the dwellings not listed in the directory, the sample might be drawn by first obtaining a sample of blocks, and making a field check of the listings in the directory for the sample of blocks. Such a check will indicate the particular units in these sample blocks that are not shown in the directory. Then the households found in the sample blocks and not in the directory are included in the sample. The results from the two samples, then, could be added together and the percentages and averages desired could be computed.

CONCLUSION

Heretofore, an attempt has been made to point out that there are many ways one can go about designing a sample. There are many other principles that have not been mentioned that can be used to increase the efficiency of a sample. Stratification is desirable and almost universally used, although not absolutely essential as is often thought to be the case. Alternative methods of estimation may be available, etc. As already indicated, sampling theory guides in the choice of efficient methods from among the available alternatives.

The same fundamental principles are applicable in problems of sampling business establishments, dwelling units, farms, factories, and other groups, although the relative importance of the particular principles to be used varies among the different problems. One can make variations in design to meet various administrative limitations and conditions and to make the maximum use of the particular resources available. The practical job of sampling design involves the use of sampling theory, a knowledge of the techniques of collection and enumeration, and a thorough search for the best resources available, and the use of these jointly in order to get the most reliable results per unit of cost. There is no *one* way that a probability sample need be drawn. It can be adapted to meet the requirements of a particular situation.

It needs to be strongly emphasized that, if results of known sampling reliability are desired, it is not sufficient to designate a good sampling method. It is essential that it be carried through in both the office and field according to specifications. Therefore it must involve processes that can be carried through by the kind of personnel available. A supervisory and control organization must be established sufficiently adequate to insure that the work is done substantially as specified. This is of extremely great importance when high precision from sample results is desired, and accounts for a significant part of the cost in many sample surveys, especially where the sample is spread over a large area.

It was previously indicated that the effective use of available resources is the way to maximize the information per dollar. For practical survey designs for market research, much of the material that might be needed and useful is already publicly available. Census data published for individual blocks may be particularly useful. But it is also true that the Bureau of the Census has developed effective unpublished materials for use in its own sampling. Sometimes these materials can also be quite effective for private groups, and the Bureau of the Census is glad to make them available at cost. There is available, for example, a set of maps outlining detailed rural areas that can be thought of as blocks, that the Bureau of the Census developed jointly with the Bureau of Agricultural Economics and the Statistical Laboratory of Iowa State College. These maps can be used in drawing rural samples of the population. In urban areas, certain materials have been developed on block designation and size that sometimes can be useful in increasing the efficiency of sampling. In connection with business establishment sampling, there is available rough identification of the number of stores by blocks for most blocks in cities of 25,000 population and more. These materials are widely used in our own sampling work. They are particularly important to the census for large scale operations for the estimation of totals as well as ratios and averages, and where results of comparatively high precision may be required. The Bureau of the Census is glad to be of any assistance in exploring the possibility and desirability of using these materials for any particular job.

Some Applications Of Statistics For Auditing

JOHN NETER

John Neter is Professor of Business Administration at the University of Minnesota. This article appeared in the *Journal of the American Statistical Association* in 1952.

In auditing, extensive use is made of samples; but in basing decisions on these samples, little if any use is made of statistical techniques. In other areas of accounting, however, the application of sound statistical techniques to the interpretation of sample data is becoming increasingly frequent. The purpose of this paper is to describe a number of applications of statistics in the area of accounting, which are particularly relevant to the problems encountered in auditing.

Auditing consists of the examination of accounting records, vouchers, and other financial and legal records and documents of an organization to ascertain the accuracy and integrity of the accounting, in particular as it is reflected in the statements of financial condition and of income. The examination may be performed for an organization by its own employees or by independent public accountants. In either event, heavy reliance is usually placed on sampling or test-checking techniques. Accounts receivable are verified by circularizing a selected number of them; inventories are generally test-checked; vouchers of cash disbursements may be examined, not for the entire accounting period, but only for a portion thereof. In view of the extensive use of sampling, it is surprising indeed that auditors have seldom employed statistical techniques to help reach conclusions about the state of the accounting records.

A consequence of the lack of application of statistical techniques is that standards of sound auditing procedures are primarily subjective. The auditor usually has no objective criterion, for example, as to how much test-checking is enough. It may well be that on the whole too much sampling is being carried on today by auditors. This writing all suffers, however, from the fact that no actual applications have been studied in order to learn answers to such questions as these: What kinds of problems are likely to be encountered? What particular statistical techniques are most suitable for this area of application? What levels of risk would be economical as well as adequate?

Statistical techniques have been applied to at least three areas of relevance to auditing. These are:

1. control of clerical accuracy,
2. sampling accounting records,
3. sampling physical property.

Methods of controlling clerical accuracy *as the work is being performed* are of considerable significance to the auditor because of his interest in the maintenance of a fairly high level of clerical accuracy in the accounting records. Statistical methods of controlling accuracy are particularly relevant to this interest.

Sampling accounting records to obtain an estimate of a certain characteristic is a common auditing procedure. For example, the auditor may sample payroll records in order to determine the extent of inaccuracies in the past year's vouchers. Other information, not now generally obtained by sampling, such as the age distribution of accounts receivable that are not recorded on punch cards, might be obtained by sampling the accounting records.

Sampling physical property occurs rather often in the verification of inventory by the auditor, less often in the verification of physical plant.

CONTROL OF CLERICAL ACCURACY

One of the earliest uses of statistical techniques in controlling clerical accuracy was made in the Census Bureau. Deming and Geoffrey [1] report that sampling verification was used in the coding and punching of population and housing data for the 1940 census, in those stages where exact conformity with the enumeration was not required.

Every 20th card in a housing census folio, which includes the schedules assigned to 4 enumerators, was verified after a randomized start if the puncher's accuracy qualified him for sampling verification. Otherwise, 100 percent checking was employed. The random start was designated to each verifier daily so that neither the puncher nor the verifier knew it in advance. In order that sampling verification might be confined to operators with reliable performance, tolerance limits were designated as follows, after considering the level of errors that could be allowed and the proportion of punchers who would qualify at any given level:

To qualify: "At least two of the last four weeks must show an average error rate of not more than 1 wrong card per 100 cards punched, and no week of the last four shall show an average of more than 2 wrong cards per 100 cards punched. (Weeks during which fewer than 2000 cards were punched will not be counted.) In addition to the above, only one of the last four weeks may include a folio for which there were more than 3 wrong cards per 100 cards punched. (Folios of fewer than 300 cards will not be counted.)"

To disqualify: "A puncher will be dropped from sample verification if the average error rate for any week, determined from samples of her work, exceeds 3 wrong cards per 100 cards punched, or if it exceeds 2 wrong cards per 100 cards punched for each of two weeks out of the last four."

For administrative convenience, an error in the punching operation was defined as a card with one or more incorrect

---

[1] Deming, W. Edwards and Geoffrey, Leon, "On sample inspection in the processing of census returns," *Journal of the American Statistical Association,* 36 (1941), 351–60.

punches. A logical alternative definition would have been an incorrect punch. The total number of punchings is usually not available in cases such as this, however, and relating the number of incorrect punchings to the cards punched would somewhat complicate the mathematical model.

Savings in direct labor cost due to substituting the sampling procedure for 100 percent checking were reported as $263,000 up to the time the paper was written. Indirect savings alone covered the cost of administering the sampling plan.

Individual control charts for each puncher who was within the administrative tolerance limits were used to discover the causes of excessive error rates. These causes ranged from illegible folios to recent illness of the puncher. It is important to remember that sampling inspection was not applied to a puncher's work until he had given evidence of continuous high-quality performance. Thus the chief function of the sampling procedure was to determine whether the high-quality performance was continuing. If it was not, 100 percent checking would again be used.

Ballowe [2] reports that Alden's, Inc., a mail order business, began in the spring of 1945 to apply control chart techniques to filling customers' orders in one of the merchandise departments. After an item ordered by a customer has been picked from stock on the basis of its catalog number, size, color, and quantity, it is checked and placed on a conveyor belt to a gravity chute. At the chute, 100 work units of merchandise are selected at random several times a day and inspected. A certain number of error possibilities, including catalog number, size, color, quantity, or price, were set up for each work unit. A work unit is considered to have been incorrectly handled if an error is made in any of the specified error possibilities. The error rate is posted on control charts as quickly as possible and remedial action instituted when out-of-control points appear.

Before control charts were used in this particular merchandise department, the error rate was 3 per 100 work units. Within two weeks after the introduction of statistical control techniques, the

[2] Ballowe, James M., "Statistical Quality Control of Clerical and Manual Operations," Reprint No. 10, *Fifth Midwest Quality Control Conference,* 1950.

error rate dropped to 1.65 percent, and at the time of writing in 1949 was about 0.7 percent. In the period from January 1946 to December 1949, the error rate for all merchandise departments was reduced by 58 percent.

In the fall of 1945, similar control chart techniques were introduced at Alden's in the general offices. Among the operations put under control were the following:

1. Open envelope, remove contents, verify remittance, apply each impression to order blank.
2. Read order to see whether any phase of transaction will not be handled in regular mail order process. If so, apply special rubber stamp, making abstracts on special requests, inquiries, and complaints.

Here also, a tremendous reduction in the error rate was achieved during 1946.

More recently, the credit department at Alden's introduced control chart techniques in posting-checking operations, credit approval, follow-up typing, and related activities. Filing, for example, is sampled by collecting duplicate stencil impressions, showing the customer's name and address, at the time the work is assigned to a clerk. After the papers are filed, a sample of 100 names is selected from the duplicate stencil impressions, and the files are examined to determine whether the items selected were filed correctly. In the credit department, as in the other departments, substantial reductions in error rates were achieved by the application of statistical control techniques.

For convenience, all sample sizes at Alden's are 100 work units. Up to 6 samples may be taken in a day. While the control charts are kept by departments and thus include the work of a number of employees, records are kept of the number and types of errors of each employee for use in corrective action. Since the most effective way to eliminate errors is to make them impossible, Alden's management attempts to do this as soon as error conditions are found to exist. For example, it may be found that the transcription of a figure, during which a transposition can occur, is not at all necessary. Only if the error possibility can not be removed is the emphasis turned to the reduction of the error ratio.

The Illinois Bell Telephone Company installed as early as July 1945 a group-sequential sampling plan for verifying clerical work in the accounting department. One of these applications consisted of verifying the punching of Social Security numbers on tabulating machine cards. The purpose of this sampling plan was not primarily to detect and correct errors in the work already completed, but rather to minimize errors in the work currently being performed by promptly revealing conditions requiring remedial action. An error was defined as the incorrect punching of one or more digits of the Social Security number on the tabulating machine card. Samples were selected from the work of each individual when this was possible. For practical purposes, a consecutively performed segment of work was treated as if it were a random sample.

A group-sequential sampling plan was used systematically and continuously. The acceptable error rate was set at 0.3 percent, the unacceptable error rate at 0.9 percent. The maximum risk of accepting unsatisfactory work was placed at 10 percent, of rejecting satisfactory work at 1 percent. The size of each sample group was determined by requiring that the acceptance and rejection numbers for cumulative samples increase by unity for each successive group. This requirement makes the administration of the plan more simple and also assures that the risks (or OC curve) for grouped sampling are the same or almost the same as for sampling by individual units. Actually, the size of the group sample was slightly rounded from the theoretical size to a more convenient number. The continuous application of this group-sequential sampling plan was specified as follows: A sample of the first work is verified at once. If the work is acceptable, the next sample is taken and verified two hours later. If the work is still acceptable, the next verification is made after one day has elapsed; and similarly, the succeeding verifications are made one week and finally one month later. Rejection of the work leads to remedial action, and the verification interval reverts back to the beginning of the sequence.

Whether work already performed is to be verified if the sample verification leads to rejection depends upon the seriousness of not finding errors. The main objective here was appropriate remedial action. A distinction was made between systematic errors, likely to be due to faulty instruction, and accidental errors, likely to be the result of poor working conditions, poorly designed working papers, illness, fatigue, inexperience, and other similar factors. The nature of the remedial action depends, of course, on the cause of the errors.

The Bell System has carried on extensive tests to determine the applicability of sequential sampling plans for controlling the quality of clerical work, particularly that involved in rating toll tickets, that is, pricing long distance calls. The results of these tests have been very satisfactory. In one of a series of tests in 1948, for example, the plan not only provided for control of clerical accuracy by remedial action when necessary, but it also located 56 percent of all errors made during this time by sampling only 12 percent of the work. On the basis of this experience, certain guides for setting up sampling plans for this type of work have been established:

1. In general, the work of an individual employee should constitute a universe.

2. Setting the unacceptable quality level three times as high as the acceptable quality level and specifying the maximum risks of accepting unsatisfactory work and of rejecting satisfactory work at 0.10 provides for economical average sample sizes. The acceptable quality level should be set so that a substantial portion of employees can meet this requirement. In this connection, control charts might be used to advantage before making a decision as to what quality level can be reached by most employees suited for the work.

3. If the size of the sample group is such that acceptance and rejection numbers increase by unity, administration of the plan is facilitated.

4. A system of verification intervals, such as that suggested by Jones, should be established. The intervals may be specified in terms of time, number of assignments, or a combination of the two. The exact intervals to be used will depend on the nature of the work examined, the degree of control to be exercised, and the time and money available for sampling verification.

Single-sampling plans are also being used in the Bell System to verify the clerical work involved in rack sorting of tickets, which is the sorting of toll tickets by the two right-hand or left-hand digits of the telephone number. The purpose of this verification is not to replace 100 percent checking but rather to provide for control over the quality of the work of the individual clerks. Nevertheless, during a test period an examination of 28 percent of the tickets sorted disclosed 64 percent of all the errors made. On the basis of these trials, a number of useful guides were found:

1. For this type of verification, single sampling appears to be the only practical type, as the taking of a sample is an intricate process.

2. If the sample leads to the conclusion that the work is unsatisfactory, the remainder of the sorted tickets should be examined.

3. A missorted ticket constitutes an error.

4. For this application it seems reasonable to design the plan so that the errors in the work subsequent to sample verification will not exceed 0.5 per 1000 tickets on the average.

5. Single sampling seems to be practical in this particular case only if the assignment consists of at least 1,000 tickets and if all pockets in the rack can be used.

6. The sample should be selected by examining the contents of a number of pockets. The number of tickets subject to verification will determine the number of pockets to be selected. The suggestion that pockets which, on the basis of past experience, include most of the sorting errors should form part of the sample probably has the effect of reducing the maximum average percentage of error (or "average outgoing quality limit") below the stated requirement.

7. The sampling verification procedure should be applied to the work of each individual clerk, and verification intervals, similar to those previously mentioned, should be established.

8. Tables stating precisely which pockets are to be examined for various lot sizes may be prepared. One possible disadvantage of this suggestion is that the clerks might learn in advance which part of their work is to be sampled.

Many other instances of the application of statistical techniques to the control of clerical accuracy may be found in the Bell System companies which, together with a number of other firms and individuals, have pioneered in this development.

Sequential sampling plans are being used in the verification of posting entries of workmen's time from work reports to labor distribution summaries. Magruder[3] of the Chesapeake and Potomac Telephone Companies reports, among a number of recent applications, the use of a continuous sampling inspection technique in the verification of Western-Electric Company billings for items shipped direct from suppliers, and the use of sampling techniques in verifying daily work reports as to accuracy of accounting for material and labor.

The Standard Register Company applies a single-sampling plan to the control of accuracy of its sales invoices. Previous 100 percent inspection was costly and did not detect all erroneous invoices. Shartle[4] reports that the use of sampling techniques reduced by 47 percent the time spent in verifying invoices and simultaneously maintained a satisfactory accuracy level. Samples are selected by a subjective random procedure from every group of invoices processed. Each group contains the work of several clerks. While sampling is not applied to the work of each clerk, a record of errors, by frequency and type, is kept for each clerk for remedial purposes. An invoice is incorrect if it contains an erasure, strike-over, transposition of figures, omission, incorrect quantity, incorrect unit price, incorrect extension, incorrect total, and so on. For a lot of invoices to be satisfactory, at least 99.25 percent of the individual invoices must be correct; a quality level of 98 percent or less is unsatisfactory. Risks of rejecting satisfactory work and of accepting unsatisfactory work are set at 5 percent. The entire lot is verified if the number of incorrect invoices in the sample equals or exceeds the rejection number. In addition, a control chart is employed. Shartle reports that out-of-control points have brought to light conditions such as improper or nonuniform training and improper placement of personnel.

An application of the control chart technique to controlling the accuracy of recording plane reservations is reported by Brinkman[5] of United Air Lines. About 10,000 incoming messages

[3] The Chesapeake and Potomac Telephone Companies, *Summary of New Sampling Applications, October 1949 to August 1950.*

[4] Shartle, Richard B., "Quality Control in the Office," *Paperwork Simplification,* No. 16 (1949), 11–13.

[5] Brinkman, J. S., "United Air Lines Speeds Reservations," Paperwork Simplification, No. 16 (1949), 9–10. Personal communication, 1950.

representing all space transactions for the United States are received daily in Denver where space for seats on United Air Lines is controlled. The phoned messages are penciled on incoming message slips. The phone wires are tapped 3 times a day and 200 consecutive messages are recorded each time. A carbon of the message slip is then checked against the recording. A message slip may be incorrect as to flight number, date, number of seats, stations involved, or even the recording of some item completely foreign to those that entered the conversation. Three-sigma control limits are used, the average being about 99.5 correct messages per 100 transcribed. It is not considered necessary to keep control charts separately for each operator, since out-of-control points are rare, but records of errors by type and incidence are kept for each employee. An important reason for the use of control charts in this case is to convince management that the personal element in telephone communication does not preclude accurate work. The control chart technique is also applied to the transcription of the message slip data to the space control charts.

The applications cited illustrate the great diversity of circumstances in which statistical techniques have been applied successfully to the control of clerical accuracy. What are the conditions necessary to make these applications successful? Following Jones's [6] suggestions, five requirements may be listed:

1. The purpose of the application of sampling techniques should not be to discover every error made but rather to establish control over the quality of clerical work. If the detection of all errors is necessary, sampling techniques are inapplicable.

2. The work should be divisible into essentially similar units; in other words, the operation is repetitious.

3. The volume or flow of the work should be large. The reason for this requirement is that clerical work does not lend itself readily to a quantitative measurement. Usually, it must be classified dichotomously, and this makes relatively large samples necessary for reasonable protection against incorrect decisions. For these relatively large samples to be economical, it is in turn necessary that the lot size be fairly large.

[6] Jones, Howard L., "Sampling Plans for Verifying Clerical Work," *Industrial Quality Control*, 3, No. 4 (1947), 5–11.

4. Incorrect units should be clearly defined, so that each unit of work can be classified readily as correct or incorrect.

5. The work should be completed at fairly frequent intervals. This permits frequent sampling, and thus remedial action can be taken promptly when it is necessary.

A number of further conclusions can be drawn from the examples presented here:

6. The auditor's best assurance that the clerical accuracy of the accounting records is reasonably satisfactory after his examination is that statistical control over clerical accuracy was exercised in the first place. He should therefore encourage, wherever possible, the use of statistical techniques to control accuracy as clerical work is done.

7. The plans so far developed generally have utilized either control chart techniques or the continuous, systematic application of acceptance sampling plans. The latter sometimes involves 100 percent verification if the number of errors in the sample equals or exceeds the rejection number, but not always. At any rate, little attempt seems to have been made so far to use continuous sampling plans, which provide protection relating to the entire process. While it is true that these continuous plans involve a considerable amount of record-keeping and require that labor for verification be available on a demand basis, it may be possible to develop other plans that overcome these objections and still provide protection relating to the entire process. This would appear to be a most desirable step because routine clerical work does generally constitute a continuous process. Also, it may be possible to reduce inspection costs while still maintaining required protection because the frequency of sampling in continuous plans will be governed by the quality of past performance. A major step in this direction seems to be the various verification intervals used, for instance, in the Bell System, which take past quality performance into account in determining the frequency and extent of sampling.

In that connection, Jones has pointed out that with short verification intervals, which is assumed to be the equivalent of sampling the same infinite lot a number of times, the risk of accepting the work each time is less than the specified risk for a single sample. Hence he suggests that, when short verification intervals are used, the risk of accepting unsatisfactory work specified for the sampling plan may be increased somewhat. It may be added that the risk of rejecting satisfactory work at least once on the basis of several samples is greater than the risk of rejecting it on the basis of a single sample. Thus, this risk for a single sample should be made rather small for short verification intervals. The risk of rejecting satisfactory work was actually specified to be only 0.01 in the example that Jones discusses.

8. On the basis of the variety of cases in which control over clerical accuracy was successfully achieved by applying statistical sampling techniques. it is reasonable to conclude that many further types of situations, both in business and in government, can be handled successfully by statistical control techniques.

SAMPLING ACCOUNTING RECORDS

Cases available in this area are rather scarce, the application of statistical sampling techniques to accounting records being quite new. The auditor generally samples accounting records in order to determine the correctness of the recording of a transaction. Although the sampling plans described in this section have other objectives, they can nevertheless be adapted to the auditor's purpose.

Magruder [7] has reported an interesting application made by the Chesapeake and Potomac Telephone Company of Baltimore City. It is necessary to ascertain periodically the distribution of telephones by type of apparatus, of which there are six. The plant department maintains records showing the type of apparatus at each customer location. While a complete inventory of these records could be taken, samples provide the information more quickly and cheaply, and substantially as accurately.

First the universe of telephones was divided into three strata:

1. Dial offices, where a subscriber line card shows the number of telephones by type of apparatus for each telephone number.
2. Nondial offices, where a subscriber line card shows the total number of telephones by type of apparatus for all the customers on 1-party, 2-party, 4-party, and rural lines.

[7] *Sample Design—Reconciliation of Continuing Property Records: Station Apparatus Account,* 1950.

3. Private branch exchanges (PBX's), where records of the number of telephones by type of apparatus for each PBX extension line are generally located.

The subscriber line card or the PBX extension line was chosen as the sampling unit, since the universe was already enumerated in this manner. Sample sizes were then determined by imposing certain precision requirements which will not be discussed here. The selection of the sample proceeded as follows:

1. In dial offices, every 144th subscriber line card was drawn, starting with a randomly chosen line number for each office.
2. In manual offices, each 96th subscriber line card was drawn, starting with a randomly chosen line number for each office.
3. The sample of PBX extension lines was chosen in three stages, using sampling with probability proportionate to size, simple random sampling, and systematic sampling, successively.

For each subscriber line card or PBX extension line selected, information as to the number of telephones by type of apparatus was then obtained and by appropriate techniques combined into population estimates.

A method of evaluating the precision of the sample was incorporated into the sample design. It consists of the use of subsamples, originally suggested by Tukey.[8] If a sufficient number of adequately large independent subsamples is used, each covering the entire universe, information may be obtained from them as to the precision of the subsample results, even though the selection of each subsample was not random.

To obtain the information as to the distribution of telephones by type of apparatus by complete examination would constitute a rather costly job. The following statement by Magruder is, therefore, especially significant: "The reconciliation of plant quantities with the accounting records is one field where major savings are in prospect by the use of sampling. We have done enough sampling in this field to feel definitely assured of success. Continued research is necessary, however, to reach sample designs of improved efficiency. This involves the usual problems of definition of sample unit, possibility of stratification, selection of

[8] Deming, William Edwards, *Some Theory of Sampling*. New York: John Wiley & Sons, Inc., 1950, p. 96.

sample elements and precision computations."

The Chesapeake and Potomac Telephone Company has applied statistical sampling techniques to other accounting records in order, for example, to obtain a distribution of disconnected equipment by age bands, to audit the classification of troubles reported by subscribers, and to segregate the book cost of outside plant according to its usage for local, state, or interstate business. The sample estimate of the proportion of plant devoted to interstate business, for instance, had a margin of uncertainty of 1.6 percent at the two-sigma level and was obtained at only about one-tenth of the cost of a complete survey. Furthermore, Magruder declares, "the sheer size of a complete survey mitigates against intelligent and detailed scrutiny of records," which is possible when sampling is used. Therefore "we have reason to believe that the sample result is more precise than a complete survey". This aspect of sampling has also been observed in other types of sample surveys.

Jones [9] has reported an application of statistical sampling techniques to accounting records by the Illinois Bell Telephone Company. The particular information desired by the company was the mean and distribution of the number of local telephone calls, according to the various classes of service offered, as well as the mean telephone usage for all classes of service combined. This information is obtainable from the company's billing records. In setting up the sampling plan, stratification was employed both by central office areas and by classes of service. For purposes of determining optimum allocation of the sample, it was found that the standard deviations of the local message usage for each class of service are about the same for the different central office areas, but that there are important differences between classes of services. The dispersion is greater for business than for residence telephones and greater for individual than for party lines. Minimum requirements as to accuracy were set with respect to the means and distributions of the telephone usage for each class of service, as well as with respect to the mean for all classes com-

[9] Jones, Howard L. "Design of Samples for 'Within Company' Analysis and Control," *Business Application of Statistical Sampling Methods,* Proceedings of conference conducted by University of Illinois and Chicago Chapter, American Statistical Association, May 1950.

bined. Selection of the sample was performed by mentally dividing a given file into as many more or less equal parts as the number of cards to be selected from it, and then picking a card from each part in haphazard fashion. This is somewhat of a systematic sample, but it has been found that the sample means in this application appear to be distributed about the same as the means of random samples. Incidentally, a sampling interval of 100 would be poor for listings of telephone numbers, since an unusually large proportion of customers with heavy usage have telephone numbers ending in even hundreds. Unused telephone numbers are distributed irregularly in the actual card file, so these constitute less of a problem than if some other record were used in making the sample selection. All customers with private switchboards were included in the sample because their number is relatively small and the distribution of usage is severely skewed to the right; similarly all customers with more than one line whose usage exceeds 5000 units during the month were included. Methods of improving the randomization of the samples for the remaining customers are being considered.

Sampling of accounting records has also been undertaken by governmental agencies. The Bureau of Old-Age and Survivors Insurance of the Social Security Administration for 12 years has been sampling its universe of account numbers, which is approaching 100 million, in order to obtain up-to-date information on the characteristics of the insured population and the operations of its program. The Bureau's experience indicates that the most feasible type of sampling in this case is digital sampling; that is, selecting all accounts that have a certain digit or digits in given locations of the serial number. The device of maintaining a sample of sufficient size for tabulating detailed data and using smaller subsamples from this larger one for tabulating other data has proven itself flexible and economical.

Another instance which may be cited in this trend to sampling accounting records is the Bureau of Public Assistance's suggestion to states with large caseloads of old-age assistance, for example, to sample their accounting records in order to obtain an estimate of the distribution of assistance payments by amounts.

These examples lead to the following general observations:

1. In each case cited, the body of accounting records sampled was large. If the universe of accounting records were small, the application of sampling techniques designed to achieve a reasonable assurance of accuracy would probably be uneconomical.

2. In each example, the sample was of a recurring nature. Hence, if the changes in the universe of the accounting records were gradual over time, a sampling plan could be used repeatedly with periodic modifications.

3. The physical state of the accounting records should be such that a sample can be selected with relative ease. Records in card files, whether use of consecutive numbering is made or not, were suitable in the cases cited.

4. It is desirable to know some universe characteristics that could be estimated from the sample, and to compare the sample estimate with the known value. In the Chesapeake and Potomac sample, for instance, if the ratio of residence telephone stations to total telephone stations is known, one can compare it with the proportion of residence stations to total stations in the sample. Actually a great many such comparisons were made to provide additional assurances of the representativeness of the sample.

5. The auditor often encounters bodies of accounting records that are large. His sampling is usually of a recurring nature. Hence it would appear that under those circumstances the establishment of statistical sampling plans would be desirable. The characteristics that he would study, generally those which pertain to the accuracy of the recording of a transaction, would not be the same as the characteristics verified in the examples discussed. Nevertheless, he would encounter the same problems as to choice of sampling units, stratifications to be employed, selection of the sample and size of the sample as have been met and adequately solved in the cases cited. While, undoubtedly, many unique problems will confront the auditor in his particular applications of statistical sampling techniques to accounting records, the experience from the examples given should at the least encourage him to experiment with the application of statistical techniques to accounting records.

6. These successful applications of statistical sampling tech-

niques to accounting records seem to have significance far beyond the auditor. In both government and business, decisions have to be made quickly on the basis of information contained in voluminous records. Sampling often can provide such information quickly and to the necessary accuracy at reasonable cost. Certainly the lack of complete accuracy of sampling results should not prevent the application of these techniques. The cases cited previously are by no means the only ones in which statistical sampling techniques have been applied to accounting records. Applications made so far, nevertheless, represent only a small beginning on a large field of potential applications—a field that will probably be developed rather rapidly with the need for quick decisions and lack of manpower in business and in government today.

SAMPLING PHYSICAL PROPERTY

The auditor often samples inventories to verify the quantity and to ascertain the quality condition of the items. Plant and equipment is sampled more rarely by him. The experience so far obtained from the application of statistical sampling techniques to physical property should be of great interest to the auditor.

Several companies in the Bell System have used statistical sampling techniques in order to determine the current average physical condition of their telephone plant, which consists of a wide array of distinct classes ranging from central office equipment and trucks to aerial and underground cable. The information was needed by the regulatory commissions for rate-making purposes. Jones and Magruder have reported on these applications; the latter will be cited here chiefly, although the problems encountered and the methods of solution in the two instances were similar.

The first problem to be faced in determining the physical condition of the Company's property was the method of specifying the state of physical deterioration. It was found that under field-inspection conditions, a maximum of five condition grades was practicable. They were defined as shown in the following table:

Condition Grade	Physical Condition	Percent Value (Illustrative)
A	New	95
B	Good	75
C	Fair	50
D	Poor	25
E	Worn out	10

The percent values reflecting the extent of relative physical deterioration were set on the basis of knowledge and experience combined with judgment, and may vary from one class of property to another. As a practical matter, it was found to be best to define grades A, B, C, D, and E on the basis of easily distinguished physical characteristics, and then to assign percent values to each condition grade, basing these largely on the age bands which correspond to each defined condition-grade.

In order that the sample results possess any degree of validity, uniformity of judgment on the part of inspectors is essential. It has been found that by a thorough training of the smallest practicable number of inspectors this uniformity of judgment can be achieved. Preliminary evidence indicates that on the average 8 out of 10 inspectors, after adequate training, will classify a given item in the same grade, the remaining two inspections splitting evenly between one grade higher and one grade lower. This split, because of its symmetry, does not affect the average physical condition reported.

The importance of human errors in a survey of this nature cannot be stressed enough. In any survey there are a number of sources of error, among which are the sampling errors. In this particular application, human errors are an especially important source of error unless the inspectors are first trained to develop uniformity in classifying the same physical property into the same condition grade. Furthermore, the inspectors must not be required to examine so many units of property that they cannot adequately examine each unit to be inspected. Here, then, is a case that requires the economic balancing of sampling and human errors in order to get the most reliable survey results for a given budget expenditure. Deming [10] states that an accurate de-

[10] Deming, W. Edwards, Personal Communication, 1951.

termination of the current average physical condition of plant in this case can only be carried out by sampling methods, using very small samples. Larger samples would involve human errors far outweighing the sampling errors that were encountered in the inspections which are being reported here.

The precision of a sample required for submission to a public service commission is probably greater than that necessary for other purposes. Even so, a precision range of less than ± 1.0 percent for the average percent condition of the property as a whole, with a 99.5 percent assurance, has been found practicable. Sample sizes were determined for each class of property on the basis of required precision. The sampling unit was chosen so that it can readily be enumerated from the property records and its location can definitely be determined from the information on the record. Furthermore, to the extent practicable, it is a ,unit that draws in other classes of property or that constitutes a relatively large part of the investment in its particular account. The unit "pole location" draws in not only the pole itself, but such other property items as crossarms, anchors, aerial cable, and cable terminals.

It was found that the method of selecting sampling units that best met the requirements of a property valuation and which also was easy to apply was systematic subsampling. To sample pole locations, for example, 10 independent subsamples were used.

Assignment of the inspection work was so arranged that each inspector would contribute about the same number of inspections to each of the 10 subsamples, covering every section of the area. In that way, the sample design provided not only a measure of the precision of the subsamples by means of the 10 independent subsamples, but it also provided evidence on the uniformity of judgment of the inspectors. Analysis of variance techniques were applied in testing these various aspects.

After the sampling units had been inspected, the average percent condition for each class of property was computed. These averages were then combined into the average percent condition of all property by using the amount of investment for each class as a weight.

Magruder [11] has reported another application of statistical sampling techniques to physical property which will only be mentioned here. Part of the telephone companies' coin box revenue for each month is uncollected at the end of the month, being in the coin boxes. To determine the amount of such revenue, a sample of coin boxes is taken by the Chesapeake and Potomac Telephone Companies each month.

Deming [12] presented a number of applications of sampling physical materials at the meeting of the International Statistical Institute in 1949. Lots of refined sugar imported into the United States have been sampled in order to estimate the quantity of sugar in the entire lot. Two-stage sampling has been applied to sample lots of domestic wool stored in warehouses in order to determine the percentage of clean wool in the lot. The primary unit was the bale, the secondary unit the core from the bale. By considering the cost of moving the bale into position and the cost of boring the bale, sample sizes for primary and secondary sampling units were determined to yield estimates of required precision at the most economical cost possible under the conditions.

In his recent book Deming [13] devotes a chapter to the quarterly taking of inventories of tires held by dealers registered with the Office of Price Administration. Again, problems of stratification, optimum allocation of sample to achieve the required precision with a minimum sample size, and method of selecting the sample units arose. In addition to these, the problem of nonresponse entered the picture, a problem not present in any of the cases cited previously. That a serious problem may arise when nonresponse is possible may be seen from the fact that the average tires per dealer was 50 percent higher in both December 1944 and March 1945 for the sample of nonrespondent dealers of September 1944 than the average inventory for all other dealers. This illustration serves to point out that one cannot simply

[11] See Footnote 3, p. 138.
[12] Deming, W. Edwards, "On the Sampling of Physical Materials," paper presented at the meeting of the *International Statistical Institute* held in Berne, 1949 (mimeographed).
[13] Deming, William Edwards, *Some Theory of Sampling.* New York: John Wiley & Sons, Inc. 1950, Chapter 11.

assume without some evidence that the nonrespondents are similar to the respondents.

These examples permit a few conclusions:

1. The techniques applied to sampling of physical property were the same statistical techniques applied to other types of sample surveys.

2. Special problems may be encountered, such as the method of selecting the sample or defining the physical condition of property. Special problems, however, are always present in statistical work.

3. The fact that statistical sampling techniques have been applied successfully to problems ranging from a nationwide inventory of tires held by dealers to the evaluation of the physical condition of the property of a large telephone company indicates that the auditor could have a useful tool for his reconciliation of inventory and plant and equipment to the accounting records. To make practicable use of statistical sampling techniques, the magnitude of the inventory and plant and equipment in terms of sampling units will have to be fairly large. Thus the auditor should experiment at first with his larger clients in applying statistical sampling techniques to physical property as well as to accounting records. Afterwards, the auditor may extend these activities to smaller concerns to determine at what size or level the application of statistical sampling techniques becomes uneconomical.

4. One of the cases illustrates the very basic problem of human errors in sample surveys. Such errors are important in auditing. The auditor, therefore, must study the relative magnitudes of human and sampling errors in the audit of physical property as well as of accounting records so that an economic balance between the two can be reached. In that connection, he must remember that a more thorough and efficient scrutiny of property or records is possible when the number of items to be scrutinized is small than when it is large.

5. Business and government could probably make more extensive use of the sampling of physical property in order to have quick and ready information on which decisions and control may be based.

Tests of Hypotheses Regarding Bilateral Monopoly

LAWRENCE FOURAKER and STANLEY SIEGEL

Lawrence Fouraker is Professor of Business Administration at Harvard University. The late Stanley Siegel taught at Pennsylvania State University. This article comes from their book, *Bargaining and Group Decision Making*, published in 1960.

1. THE HYPOTHESIS CONCERNING THE PARETIAN OPTIMA

Introduction

A theoretical model exists that offers a solution to the bargaining situation in which both bargainers are unique, that is, a bilateral monopoly bargaining situation. An example of this situation is presented by the single buyer of a commodity with no close substitutes negotiating with the only seller of that commodity.

One of the predictions yielded by the theoretical model is that the contracts arrived at in bargaining in this situation would tend to the output that maximizes joint profit, i.e., would tend to the *Paretian optima*. It was shown that if

A = the price axis intercept of the average revenue function
A' = the price axis intercept of the average cost function
B = the slope (negative) of the average revenue function
B' = the slope of the average cost function
Q = quantity

then the quantity that maximizes joint profits Q_m is

$$Q_m = \frac{A - A'}{2B + 2B'}$$

This paper presents the experiments which were designed to test the prediction that bilateral monopoly contracts would tend to the Paretian optima, that is, would be reached at or near the quantity Q_m that maximizes joint payoff. The first experiment to be reported was directed solely toward testing this hypothesis.

The Experimental Test

Subjects and procedure. Twenty-four male undergraduate students (12 bargaining pairs) participated in this experiment (experimental session 1). Each subject was given a set of iso-profit tables appropriate to his role (buyer or seller). Buyers and sellers were instructed separately, and then taken individually to cubicles where they were isolated from all but the experi-

TABLE 1
QUANTITY AND JOINT PAYOFF AGREED UPON IN CONTRACTS
REACHED IN EXPERIMENTAL SESSION 1

Quantity	Joint payoff
6	$ 9.60
8	10.50
9	10.80
9	10.80
9	10.80
9	10.80
10	10.70
10	10.70
10	10.70
10	10.70
15	6.90

menters and their assistants. Negotiations were conducted in silence, using written offers and counter-offers transmitted by the research personnel.

The iso-profit tables used in Experimental Session 1 were derived from the following set of parameters: $A = \$2.40$, $A' = \$0.00$, $B = \$0.033$, and $B' = \$0.10$. Thus, as is shown by the vertical line in Fig. 1, the Paretian optima fall at $Q_m = 9$, and the maximum possible joint payoff is \$10.80.

Results. Table 1 contains the observations on 11 bargaining pairs, one pair of the original 12 having failed to come to any agreement within the time allowed (two hours). Shown in the table is the quantity arrived at in the contract, and the joint pay-

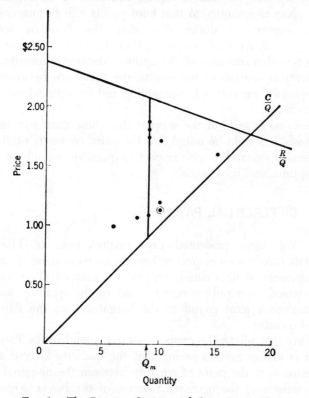

FIG. 1. The Paretian Optima and Contracts in Experimental Session 1. The Encircled Dot Represents Two Identical Observations.

off contingent on that quantity. In Fig. 1, these results are shown graphically. The Paretian optima are represented by the heavy vertical line. The 11 observations are shown as dots, with an encircled dot representing two identical observations.

The mean quantity arrived at in Session 1 is $\overline{Q} = 9.54$. The difference between this observed mean and that expected ($Q_m = 9.00$) is insignificant: $t = 0.84$, $.50 > p > .40$.

Discussion

The data tend to support the hypothesis regarding the Paretian optima, i.e., that contracts will tend to be negotiated with respect to quantity so that joint profits will be maximized. In the experiment under discussion, the Paretian optima fell at $Q_m = 9$. As Table 1 reveals, 9 of the 11 teams negotiated contracts within one unit of the optima. Moreover, according to the statistical analysis of the results, the difference between the mean quantity arrived at in bargaining and the optimal quantity is insignificant.

However, in spite of the support that these data give to the hypothesis, it should be noted that the pairs' contracts exhibited considerable variability with respect to quantity around the Paretian optima, as Fig. 1 reveals.

2.   DIFFERENTIAL PAYOFF

We have presented experimental tests of bilateral monopoly theory with respect to hypotheses concerning quantity. The experimental data clearly support the theoretical contention that bilateral monopoly contracts tend to the quantity output that maximizes joint payoff to the bargainers, to the Paretian optimal quantity.

To say that bilateral monopoly contracts tend to the Paretian optima is tantamount to saying that the *quantity* arrived at in bargaining is at the point of equality between the marginal cost of the seller and the marginal revenue of the buyer. Figure 2 shows an average cost and an average revenue function and their associated marginal cost and marginal revenue functions.

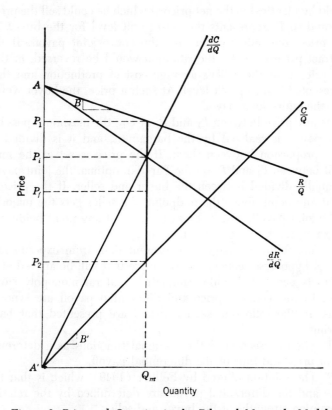

Figure 2. Price and Quantity in the Bilateral Monopoly Model

The marginal cost function dC/dQ intersects the marginal revenue function dR/dQ at Q_m, the quantity that maximizes joint payoff. Thus, if either bargainer were to move from the quantity designated by the intersection of the marginal functions to some other quantity, he could maintain his previous profit level only if his rival's profit were reduced.

Inspection of Fig. 2 will reveal that, although *quantity* is determined at Q_m, the *price* at which the quantity may be exchanged can lie anywhere between P_1 and P_2. These limits of price are set by the average revenue function of the buyer R/Q and the average cost function of the seller C/Q. If the contract price were P_1, the price that the buyer would pay for the product would be identical to the net price at which he could sell the product, and so P_1 represents the zero profit level for the buyer. At this price, the seller would take the entire joint payoff. If the contract price were P_2, the situation would be reversed, in that this price is at the seller's average cost of production and thus represents his zero profit level. At such a price, the buyer would take the entire joint payoff.

At any price between P_1 and P_2 at Q_m, the maximum possible joint payoff is realized by the bargainers, and it is divided in some proportion between them. If the price is set at the midpoint between P_1 and P_2 on the Paretian optima, the joint payoff is equally divided between the buyer and seller. If the price is set at any point above the midpoint, the seller gets the majority of the joint payoff, and if the price is set at any point below the midpoint, the buyer gets the majority.

In the following pages, we present data from experimental tests of hypotheses concerning the price that will be arrived at in contracts negotiated under simulated bilateral monopoly situations. In this context, price and differential payoff are synonymous. In the following sections, data are presented that have bearing on:

1. The intersection of the marginal functions as a determinant of price and thus of the differential payoff,

2. The solution offered by Fellner (1949), which is that the price and the differential payoff are determined by the relative bargaining strengths of the buyer and seller.

3. THE MARGINAL INTERSECTION HYPOTHESIS AND THE FELLNER HYPOTHESIS

INTRODUCTION

It has been suggested that, when bargainers negotiate under incomplete information, each will offer combinations of price and quantity along his own marginal function. That is, when the seller knows his own cost functions and the buyer his own revenue functions but neither has information concerning his rival's functions, the seller will offer combinations along his marginal cost function, starting with a high price and quantity and making downward concessions as negotiations require, while the buyer will offer combinations along his marginal revenue function, starting with a low price and quantity and making upward price concessions as negotiations require. Thus, equilibrium will be achieved at the intersection of the marginal functions, yielding a contract (P_i in Fig. 2) which maximizes joint payoff. The differential payoff is, under this account, a function of the relative slopes of the cost and revenue functions, B and B'.

The expected price according to the marginal intersection hypothesis, P_i, is

$$P_i = \frac{AB' + A'B}{B + B'} \tag{2}$$

where
  A = the price axis intercept of the average revenue function
  A' = the price axis intercept of the average cost function
  B = the slope (negative) of the average revenue function
  B' = the slope of the average cost function

Fellner (1949) has proposed a hypothesis that stands in contrast to the marginal intersection hypothesis. His position is that the price which will be arrived at in a bilateral monopoly situation and which will determine the division of the profits (the differential payoff) depends on the relative bargaining strengths of the buyer and seller. The price predicted under Fellner's hypothesis falls on the Paretian optima

$$A - B\,\frac{A - A'}{2B + 2B'} \geq P \geq A' + B'\,\frac{A - A'}{2B + 2B'} \tag{3}$$

and has a particular value that depends on the relative bargaining strength of the rivals in negotiation.

If the Fellner position is correct, then, if a large number of bargainers are randomly assigned to pairs and if within each pair the roles of buyer and seller are randomly assigned, so that it may be assumed that relative bargaining strength is randomly distributed among buyers and sellers, the prices arrived at in bargaining contracts may be expected to form a random symmetrical distribution over the range between average revenue and average cost. Further, under the Fellner hypothesis, it may be expected that the distribution of prices will have its central tendency at the midpoint of the Paretian optima

$$P_f = \frac{3AB' + 3A'B + AB + A'B'}{4B + 4B'} \tag{4}$$

The price at which the marginal functions intersect P_i will be different from the midpoint of the Paretian optima P_f in any situation in which the revenue and cost functions have unequal slopes, i.e., $B = B'$. Figure 2 illustrates one such situation.

Thus we have conflicting predictions concerning the price at which contracts will be negotiated in bilateral monopoly situations. The experiment to be reported allows a test of whether the data are more consistent with the marginal intersection hypothesis or the Fellner hypothesis. That is, they provide a test of the prediction that negotiated prices will tend to fall at that point which is the intersection of the functions that stand in a marginal relation to the buyer's average revenue function and the seller's average cost function, against the prediction that negotiated prices will tend to the midpoint of the Paretian optima when bargaining strength between buyers and sellers is controlled.

The Experimental Test

Subjects and procedure. In the experimental test of this hypothesis, conducted in Experimental Session 1, the subjects were 22 male undergraduates recruited from classes in elementary economics. The influence of individual differences in bargaining strength was controlled by random assignment of the following: identity of pair members, identity of buyers and sellers, identity of initiators of bargaining.

Each subject received a set of iso-profit tables, which were derived from the following parameters: $A = \$2.40$, $A' = \$0.00$, $B = \$0.033$, and $B' = \$0.10$.

With these parameters, the Fellner hypothesis is that the central tendency of prices in contracts negotiated will be $P_f = \$1.50$. The marginal intersection hypothesis is that prices will be negotiated at $P_i = \$1.80$. Observe that $P_i > P_f$ since $B' > B$: The marginal functions intersect above the midpoint of the Paretian op-

TABLE 2

CONTRACTS NEGOTIATED BY BARGAINING PAIRS
IN EXPERIMENTAL SESSION 1

Quantity	Price	Profits		
		Buyer	Seller	Joint payoff
6	$1.00	$7.20	$2.40	$ 9.60
8	1.07	8.40	2.10	10.50
9	1.10	9.00	1.80	10.80
10	1.15	9.20	1.50	10.70
10	1.15	9.20	1.50	10.70
10	1.21	8.60	2.10	10.70
15	1.62	4.20	2.70	6.90
10	1.74	3.30	7.40	10.70
9	1.77	3.00	7.80	10.80
9	1.83	2.40	8.40	10.80
9	1.90	1.80	9.00	10.80
Mean 9.54	$1.41	$6.03	$4.24	—

tima. For this set of iso-profit tables, the Paretian optimal quantity is $Q_m = 9$, and the maximum joint payoff is $10.80.

Results. Table 2 presents information on the contracts negotiated by each of the 11 bargaining pairs. This information is presented graphically in Fig. 3, with the relevant cost and revenue functions shown.

The mean price arrived at in the various contracts is $\bar{P} = \$1.41$. The deviation of this value from the price at the intersection of the marginal functions ($P_i = \$1.80$) is significant at well beyond the .01 level : $t = 3.59$, $df = 10$, $p < .005$. On the other hand, the deviation of the observed mean price from the price expected under the Fellner hypothesis ($P_f = \$1.50$) is insignificant: $t = 0.81$, $df = 10$, $.50 > p > .40$.

Discussion

On the basis of the data from Experimental Session 1, as presented in Table 2, the marginal intersection hypothesis must be rejected in favor of the Fellner hypothesis. The mean of the prices arrived at in the various contracts, $P = \$1.41$, is significantly different from $P_i = \$1.80$, but it is not significantly different from $P_f = \$1.50$. Moreover, whereas the marginal functions intersect above the midpoint of the Paretian optima and thus the

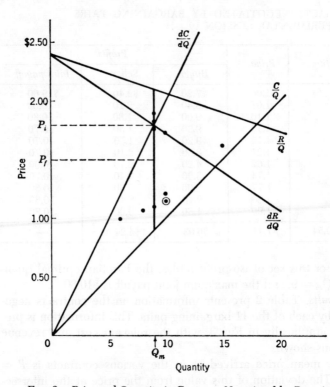

FIG. 3.   Price and Quantity in Contracts Negotiated by Bargaining Pairs in Experimental Session 1. The Encircled Dot Represents Two Identical Observations.

marginal intersection hypothesis predicts an advantage in differential payoff for the seller, in the bargaining in Session 1 the buyers did slightly better than the sellers, although the difference between the two was not statistically significant.

Gibrat's Law and the Growth of Firms

EDWIN MANSFIELD

Edwin Mansfield is Professor of Economics at the Wharton School of the University of Pennsylvania. This piece comes from his paper, "Entry, Gibrat's Law, Innovation, and the Growth of Firms," in the 1962 *American Economic Review*.

Gibrat's law is a proposition regarding the process of firm growth. According to this law, the probability of a given proportionate change in size during a specified period is the same for all firms in a given industry—regardless of their size at the beginning of the period. For example, a firm with sales of $100 million is as likely to double in size during a given period as a firm with sales of $100 thousand. Put differently, Gibrat's law states that:

$$S_{ij}^{t+\Delta} = U_{ij}(t, \Delta)S_{ij}^t \tag{1}$$

where S_{ij}^t is the size of the jth firm in the ith industry at time t, $S_{ij}^{t+\Delta}$ is its size at time $t + \Delta$, and $U_{ij}(t, \Delta)$ is a random variable distributed independently of S_{ij}^t.

Since this law is a basic ingredient in many mathematical models designed to explain the shape of the size distribution of firms, and since this law has interesting implications regarding the determinants of the amount of concentration in an industry, some importance attaches to whether or not it holds. This paper provides tests based on data for practically all firms—large and small—in three individual industries: the steel, petroleum, and rubber tire industries.

A simple way to test Gibrat's law is to classify firms by their initial size (S_{ij}^t), compute the frequency distribution of $S_{ij}^{t+\Delta}/S_{ij}^t$ within each of these classes, and use a x^2 test to determine whether the frequency distributions are the same in each class. We rely heavily on this test, but supplement it with others. The basic data used in these tests are described and presented in the Appendix.

Gibrat's law can be formulated in at least three ways, depending on the treatment of the death of firms and the comprehensiveness claimed for the law. First, one can postulate that it holds for all firms—including those that leave the industry during the period. If we regard the size (at the end of the period) of each of these departing firms as zero (or approximately zero), this version can easily be tested. The results, shown in Table 1, indicate that it generally fails to hold. In 7 of the 10 cases, the observed value of x^2 exceeds the critical limit corresponding to the .05 significance level.[1]

Why does this version of the law fail to hold? Even a quick inspection of the data shows one principal reason. The probability that a firm will die is certainly not independent of its size. In every industry and time interval, the smaller firms were more likely than the larger ones to leave the industry. For this reason (and others indicated below), this version of the law seems to be incorrect.

Second, one can postulate that the law holds for all firms other than those that leave the industry. Omitting such firms, we ran another series of x^2 tests, the results of which are shown in Table 1. In 4 of the 10 cases, the evidence seems to contradict the hypothesis, the observed value of x^2 exceeding the limit corresponding to the .05 significance level.

To see why this version must be rejected, note that Equation (1) implies that

$$\ln S_{ij}^{t+\Delta} = V_i(t, \Delta) + \ln S_{ij}^t + W_{ij}(t, \Delta) \qquad (2)$$

where $V_i(t, \Delta)$ is the mean of $\ln U_{ij}(t, \Delta)$ and $W_{ij}(t, \Delta)$ is a homoscedastic random variable with zero mean. Thus, if $\ln S_{ij}^{t+\Delta}$ is plotted against $\ln S_{ij}^t$, the data should be scattered with constant

---

[1] The size classes and the cutoff points for $S_{ij}^{t+\Delta}/S_{ij}^t$ used in these tests are described in the Appendix.

TABLE 1

OBSERVED VALUE OF χ^2 CRITERION, ESTIMATED SLOPE OF REGRESSION OF $\ln S_{ij}^{t+\Delta}$ ON $\ln S_{ij}^t$, AND RATIO OF VARIANCES OF GROWTH RATES OF LARGE AND SMALL FIRMS, STEEL, PETROLEUM, AND RUBBER TIRE INDUSTRIES, SELECTED PERIODS.[a]

Item	Steel				Petroleum				Tires	
	1916–1926	1926–1935	1935–1945	1945–1954	1921–1927	1927–1937	1937–1947	1947–1957	1937–1945	1945–1952
χ^2 criterion:										
Including deaths	9.0	17.0[b]	22.5[b]	7.8	29.2[b]	44.9[b]	25.6[b]	42.7[b]	9.3	22.9[b]
Excluding deaths	7.1	3.3	9.5[b]	3.4	2.8	22.1[b]	17.7[b]	8.9	6.3	6.6[b]
Degrees of freedom (χ^2 tests):										
Including deaths	6	6	6	6	6	6	6	6	6	4
Excluding deaths	4	4	4	4	4	4	4	4	4	2
Estimated slope:[c]										
Excluding deaths	.88[b]	.99	.92[b]	1.00	.94	.88[b]	.99	.94	.97	.97
Large firms only	.94	.96	1.00	.98	.99	.98	.93	1.10	1.07	.89
Standard error of slope:										
Excluding deaths	.05	.04	.03	.04	.05	.04	.03	.04	.05	.04
Large firms only	.16	.16	.07	.06	.24	.14	.07	.07	.10	.05
Number of firms:										
Excluding deaths	72	66	64	69	128	116	156	106	34	31
Large firms only	7	9	11	12	7	11	16	17	11	12
Ratio of variances of growth rates of large and small firms:[d]										
Excluding deaths	8.96[b]	.80	37.40[b]	5.06[b]	43.27[b]	19.25[b]	63.56[b]	147.1[b]	16.16[b]	.31
Large firms only	.63	161.00[b]	.90	8.50[b]	3.50	7.75[b]	4.00[b]	3.6[b]	39.25[b]	8.67

[a] Symbols: S_{ij}^t is the size of the jth firm in the ith industry at time t, and $S_{ij}^{t+\Delta}$ is its size at time $t+\Delta$. For the classification of firms by size and the classification of $S_{ij}^{t+\Delta}/S_{ij}^t$ used in each industry in the χ^2 tests, see the Appendix. The number of degrees of freedom equals $(a-1)(b-1)$ where a is the number of size classes and b is the number of classes of $S_{ij}^{t+\Delta}/S_{ij}^t$ in the contingency table.

[b] For χ^2 criteria and ratios of variances, this means that the probability is less than .05 that a value would be this large (or larger) if Gibrat's law held. For estimated slopes, this means that they differ significantly from unity (.05 significance level).

[c] The number of firms in each regression is shown under "Number of firms."

[d] The firms regarded as "small" and "large" in the first row are as follows: In steel, small firms have 4,000–16,000 and large firms have 256,000–4,096,000 tons of capacity. In petroleum, small firms have 500–999 and large firms have 32,000–511,999 barrels of capacity. In tires, small firms have 80–159 and large firms have 640–5,119 employees. The firms regarded as "small" and "large" in the second row are described in footnote 3.

Source: See the Appendix.

variance about a line with slope of one. Table 1 contains the least-squares estimate of the slope of each of these lines. In half of the cases where the law was rejected the slope is significantly less than one.

In addition, the variance of $S_{ij}^{t+\Delta}/S_{ij}^{t}$ tends to be inversely related to S_{ij}^{t}. Taking in each case a group of small firms and dividing the variance of their values of $S_{ij}^{t+\Delta}/S_{ij}^{t}$ by the variance among a group of large firms, we obtain the results shown in Table 1. In 8 of the 10 cases the variances differed significantly. Thus, contrary to this version of the law, smaller firms often tend to have higher and more variable growth rates than larger firms.

Third, one can postulate that the law holds only for firms exceeding the minimum efficient size in the industry—the size (assuming the long-run average cost curve is J-shaped) below which unit costs rise sharply and above which they vary only slightly. This is the version put forth by Simon and Bonini, although it seems to be a stronger assumption than they require.[2] One is faced once again with the problem of whether or not to include firms that die. We excluded them, but the major results would almost certainly have been the same if they had been included.

This version was tested in two ways. First, we estimated the slope of the regression of ln $S_{ij}^{t+\Delta}$ on ln S_{ij}^{t}, but included only those firms that were larger than Bain's estimate of the minimum efficient size. The results are quite consistent with Gibrat's law (the slopes never differing significantly from one). Second, we used F tests to determine whether the variance of $S_{ij}^{t+\Delta}/S_{ij}^{t}$ was constant among these firms. Contrary to Gibrat's law, the variance of $S_{ij}^{t+\Delta}/S_{ij}^{t}$ tends to be inversely related to S_{ij}^{t} in 6 of the 10 cases.[3]

---

[2] Herbert Simon informs me that the version of Gibrat's law they used in their paper is not required to obtain the Yule distribution and that their proof will hold if the expected value of $S^{t+\Delta}/S^{t}$ does not vary with S^{t} regardless of whether or not the variance of $S^{t+\Delta}/S^{t}$ depends on S^{t}. Our results do not contradict these weaker assumptions for firms above the minimum efficient size and consequently they do not contradict their findings based on them. But they do contradict the version of Gibrat's law in their paper.

[3] The x^2 tests had to be abandoned here because of the small number of firms. Firms with more than 64,000 barrels of capacity (petroleum), 1,000-000 net tons of capacity (steel), or .8 percent of total employment (tires) were included in the regression. The number included in each case is shown in Table 1. The fact that none of the slopes differs significantly from

Thus, regardless of which version one chooses, Gibrat's law fails to hold in more than one-half of these cases.

Appendix

In this Appendix, we describe the way in which firms were classified by S_{ij}^t and $S_{ij}^{t+\Delta}/S_{ij}^t$ in the χ^2 tests in Table 1. In the tests in which deaths were included, the following size classes were used. In steel, we classified firms by their value of S_{ij}^t into four classes: 4000–15,999 tons, 16,000–63,999 tons, 64,000–255,999 tons, and 256,000–4,096,000 tons. In tires, we used four classes: 20–79 men, 80–159 men, 160–639 men, and 640–5119 men. And in petroleum, there were four classes: 500–999 barrels, 2000–3999 barrels, 8000–15,999 barrels, and 32,000–511,999 barrels. To cut down the computations involved, only firms in these classes were included. Thus some of the largest and smallest firms were omitted in steel and tires, and some small, medium-sized, and large firms were excluded in petroleum. But had all firms been included, the results would almost certainly have been much the same.

In all cases, the firms in a size class were divided into three groups: those where $S_{ij}^{t+\Delta}/S_{ij}^t$ was less than .50, between .50 and 1.50, and 1.50 or more. These classes were chosen so that the expected number of firms in each cell of the contingency table would be five or more. (According to a well-known rule of thumb, the expected number in each cell should be this large.) This did not always turn out to be the case, but further work showed that the results would stand up if cells were combined.

one indicates that there is no evidence that among these firms the average growth rate depended on a firm's initial size.

In the variance ratio tests we divided these firms into two size (S_{ij}^t) groups, the dividing line being 150,000 barrels of capacity (petroleum), 3,000,000 tons of capacity (steel), and 30,000 employment (tires). Then F tests were used to determine whether the variances of $S_{ij}^{t+\Delta}/S_{ij}^t$ differed. This test is not too robust with regard to departures from normality, but it should perform reasonably well here.

Note that, in petroleum and tires, we include firms that are more than one-half of the minimum efficient sizes. According to Bain the cost curve is quite flat back to one-half of those sizes. Thus, it seemed acceptable to include the additional firms and to increase the power of the tests in this way.

With the following exceptions, these same classifications were used in the tests in which deaths were excluded. In steel and tires, the two smallest size classes were combined. In some cases, firms were classified into groups where $S_{ij}^{t+\Delta}/S_{ij}^{t}$ was less than 1.00, between 1.00 and 2.00, and 2.00 or more. These changes were made to meet the rule of thumb noted. Despite these changes, the expected number of firms in some cells was not quite five, but the results would not be affected if some cells were combined.

SIMPLE REGRESSION: DEMAND AND COSTS

It is no exaggeration to call regression analysis the work-horse of economic and business statistics. In most elementary courses, the treatment of regression begins with a discussion of simple regression, the case in which there is only one independent variable that is used to explain the dependent variable. An example of a simple regression is the relationship between the sales of a particular product and the level of gross national product, the product's sales being the dependent variable and gross national product being the independent variable. The purpose of the next four articles is to present some important illustrations of the use of simple regression in analyzing demand and costs.

To begin with, Robert Ferber discusses the use of regression techniques in forecasting company or industry sales.[1] As he says, "the method involves the fitting of an equation to explain the fluctuations in sales in terms of related and presumably causal

---

[1] Parts of this article (and Dean's) deal with regressions involving more than one independent variable, but they should be understandable to readers who have studied only simple regression.

variables, substituting for these variables values considered likely during the period to be forecasted, and solving for the value of sales." He discusses the selection of variables, the units of measurement, the period of observation, the evaluation of forecasts, the possibility of improving the accuracy of future forecasts based on the same equations, and other relevant considerations.

In the following paper, E. Working points out that demand curves estimated by ordinary regression techniques may not be demand curves at all, but hybrids of demand and supply curves, or even supply curves. His paper is, of course, an early—and classic—description of the identification problem. The next paper, by Joel Dean, is also very well known. Using regression techniques, he estimates the cost functions of a hosiery mill, one of his most controversial findings being that marginal cost is constant.

The final paper, by Frederick Moore, is concerned with the measurement of economies of scale. He begins by discussing the ".6 rule" used by engineers to estimate the increases in capital cost resulting from increases in capacity. Then he presents estimates of the relationship between capital cost and capacity for a large number of products in the chemical and metal industries, these estimates being derived by simple regression. If there are constant returns to scale, the parameter, b, should equal one. Where possible, he applies t tests to determine whether b differs from one.

Sales Forecasting By Correlation Techniques

ROBERT FERBER

Robert Ferber is Professor of Economics at the University of Illinois. This paper appeared in the *Journal of Marketing* in 1954.

The use of correlation methods to forecast company or industry sales has been gaining increasing popularity in recent years. Briefly stated, the method involves the fitting of an equation to explain the fluctuations in sales in terms of related and presumably causal variables, substituting for these variables values considered likely during the period to be forecasted, and solving for the value of sales. The method possesses a number of widely recognized limitations,[1] but can nevertheless be of considerable worth as a forecasting aid. This is aside from its usefulness in testing the relevance of various factors as determinants of sales and in evaluating the relative importance of each.

The technique of deriving the equation relating sales to the explanatory variables—the regression equation—is given in detail in a variety of sources,[2] and is easily mastered. It is not the pur-

[1] Particularly (a) the assumed stability of historical relationships for the future, (b) the danger that causality may run as much in one direction as in the other, and (c) the difficulty of predicting future values of the variables used to predict sales, especially when these variables are not lagged.

[2] For example, M. Ezekiel, *Methods of Correlation Analysis;* R. Ferber, *Statistical Methods in Market Research;* A. E. Waugh, *Elements of Statistical Methods.*

169

pose of this article to consider these technical details, but rather to offer some suggestions and raise a few questions regarding the specification and use of these equations for sales forecasting. Everyone recognizes that the value of such an equation can be no more than the validity of its underlying assumptions—choice of explanatory variables, units of measurement, period of observation, and so forth—though their importance tends at times to be minimized, or routinized, in the dazzle of securing coefficients precise, at least mathematically, to six, seven, or eight significant figures.

It is the purpose of this article to re-emphasize the importance of some of these basic considerations by offering suggestions—some new, some old—regarding their resolution. These suggestions are based in large part on the writer's studies of aggregate relationships which, though admittedly not bearing directly on sales forecasting, encountered similar difficulties. Most of the discussion is concerned with the specification of the regression or forecast equation. However, consideration is also given to evaluating the adequacy of forecasts derived from a regression equation, and the possibility of improving the accuracy of future forecasts based on the same equations. It should be stressed that no attempt is made to provide comprehensive coverage of the questions raised or to present an inclusive list of questions dealing with the use of correlation techniques in sales forecasting.

PREPARING AND USING THE FORECAST EQUATION

The following basic situation is assumed: An attempt is to be made to forecast company or industry sales—it does not matter which, at the outset—by deriving a regression equation between the dependent variable, sales, and one or more presumably independent variables which are believed to cause the fluctuations in sales. This is admittedly not a very precisely sketched situation, and deliberately so, as many of the questions concerning the specification of the equation and the method deriving the forecasts are to be discussed below. In particular, questions regarding the following matters will be raised: the method of fit, the method of deriving the forecast, selection of the period of observation, units of measurement for the variables, and choice of variables for inclusion in the equation. The order of presentation of these various considerations is not the order in which they may arise in practice, but is governed primarily by which considerations are most likely to affect the others.

Method of fit

Once a forecast equation has been set up, say, that sales (S) is a linear arithmetic function of national disposable income (Y) and of advertising expenditures (A),[3] that is, $S = a + bY + cA$, the problem arises of estimating the values of the parameters of the equation, a, b, and c. This is almost invariably done by the so-called least-squares method based on deriving that equation which will minimize the sum of the squares of the deviations of the observations from the fitted line, for that type of function. It involves the construction and solution of a set of "normal" equations, and is the standard method given in statistical texts for this type of work.

In recent years this method of fit has been criticized for yielding biased estimates of the parameters. The basis for this criticism is the assumption implicit in the least-squares method that the independent variables are truly independent of the dependent variable. When the dependent and one or more independent variables refer to the same unit of time, this assumption may be negated. Thus, if steel sales were related to national income, one might argue that these sales are as much a *determinant* of national income as national income is of steel sales. In such a case, it can be shown that the true regression of sales on income and other variables is derived not by the method of least squares but by means of a system of equations containing as many equations as there are interrelated variables.

A simple illustration of such an equation system for the above example would be the following equations:

steel sales $= a + b$ (national income)
national income $= c + d$ (government expenditures)
                      $+ e$ (private investment)

[3] Assuming that such expenditures are flexible.

where we assume that government expenditures and private investment are independently determined, i.e., known beforehand.

In such a case, we have two equations in two interrelated variables, sales and national income, and unbiased estimates of the parameters, *a* and *b*, are obtainable.

The solution of such equation systems can become quite involved, and perhaps for this reason, it never seems to have caught on with sales forecasters. As things have turned out, their reluctance to derive relationships by means of equation systems appears to be justified.

There are two reasons for this. One is that equation systems that have been used to provide forecasts of economic trends on a national basis have not performed very well. The second reason is that, in the case of relationships between industry or company sales and such national indicators as income, production, and population, as are so often used in sales forecasting, the direction of causation can hardly be thought to be equally strong in both directions. Even for a key industry such as steel, national income can clearly be said to affect steel production far more than steel production affects national income.

For these reasons, the continued use of the least-square, single-equation method for deriving sales forecasting equations is warranted. Sales forecasters would do well, however, to keep an eye on developments in the use of equation systems.*

Derivation of the forecast

Discussion of this question may seem, at first sight, somewhat puzzling. For once one has the equation and the values of the independent variables, what is there to deriving the forecast other than performing the necessary substitutions and solving for the sales variable? One answer is that the function may be used to estimate not the level of savings, as is the customary procedure, but rather the *change* in savings. This is accomplished in the following manner.

Suppose our forecast function is, as above:

$$S_t^e = a + bY_t + cA_t$$

where the subscript t refers to time and indicates that all varia-

---

* Important developments have occurred in the development of equations systems since this article was written. See various econometrics texts. Editor.

bles refer to the same time interval, and the superscript e indicates the value being estimated.

To predict the change in savings is to compute $S_t - S_{t-1}$. Sales in the last time period, S_{t-1}, can be expressed in the same form as current sales but with a lagged time subscript, i.e.:

$$S_{t-1}^e = a + bY_{t-1} + cA_{t-1}$$

Therefore,

$$(S_t - S_{t-1})^e = a + bY_t + cA_t - (a + bY_{t-1} + cA_{t-1}$$

so that

estimated change in sales $= b(Y_t - Y_{t-1}) + c(A_t - A_{t-1})$

The prediction of the change in sales is made by substituting the required values of Y and A in the last equation and solving for $S_t - S_{t-1}$. The forecast of the sales level in time t is then the sum of actual sales in the preceding time period and the estimated change in sales from time $t-1$ to time t. The result is not the same as that obtained by predicting the level of sales directly from the original equation.

When is one method of deriving the forecast superior to the other? That is difficult to answer, and depends in part in a particular case on a comparison of both methods over a period of time. It can be said however, that if the forecasts by the customary method exhibit a systematic bias in one particular direction —if, say, they have been overshooting the actual figure fairly consistently—use of first differences will probably improve the accuracy of the forecasts inasmuch as the benchmark for the forecast automatically becomes the preceding level of sales. (This method is especially to be preferred over the device sometimes adopted for adding or subtracting an arbitrarily selected constant to the equation to adjust for the observed bias.) On the other hand, if the forecasts have been zigzagging around the actual figures, or the latter are subject to large random variations, working in terms of the change in sales will probably produce inferior results.

It may be of interest to note that tests of the accuracy of postwar predictions of personal savings based on relationships fitted to pre-World War II data revealed the first-difference method to be generally the more accurate of the two. This was a period, however, when the prewar-fitted functions were consistently over-estimating the true figures.

The period of observation

Two conflicting impulses are likely to arise in selecting the period the data for which are to serve as the basis for fitting the forecasting function. On the one hand, there is the urge to utilize all the data available, and thereby maximize the reliability of the coefficients, at least from a purely statistical viewpoint. On the other hand, it is desirable to have a few years, or observations, left over, so that the predictive accuracy of the function can be tested in advance. Given this conflict, what is the solution?

We may begin to answer this question by citing what is not the solution; namely, to lop off arbitrarily two or three years at one end of the data, fit the function to the other observations, and test its forecasts on these remaining years. This is not the solution because the selection of a period of observation is basically a logical consideration. In other words, the period of observation should include all the years or observations for which the nature of the given relationship does not appear to have been altered, as determined by a careful study of all available evidence.

If no event is believed to have occurred during the period under consideration to alter the sales forecasting function, all of it is included in the period of observation, for there is then no logical basis for excluding any particular observations. The determination of the predictive accuracy of the fitted function in advance then becomes a matter of judgment or securing rough estimates for the variables for years outside of the period of observation.

If, however, the parameters of the sales forecasting function are believed to have been altered at one time during this period—say, by the sudden discovery of a new use for a principal product, the use being thought to be less dependent on national income than other uses for this product—the rational approach is to divide the period into two subperiods, using the time at which the change is believed to have occurred as the division point. One period is then the period of observation and the data for the other period are used as a basis for testing the predictive accuracy of the fitted function.

If the accuracy of the forecasts does not differ appreciably from that of the function's estimates in the period of observation, a basis exists for combining both subperiods into a single period of observation and recomputing the sales forecast function using all the data available. If appreciable differences in accuracy do exist between the estimates for the two subperiods, utilization of all the available data in fitting the function is not warranted and the accuracy of forecasts for future periods may be in doubt, depending on the circumstances.

All this may seem a bit complicated, so let us illustrate the various possibilities with a few examples. The basic situation, we shall assume, is that sales of industry Z are available on an annual basis for 1923–41 and 1947–53, the industry being engaged in military goods production from 1942 to 1946. A sales forecast function is desired as an aid in forecasting industry sales for 1954 and thereafter. The form of the function is:

$$\text{sales} = a + b \text{ (disposable income)}$$
$$+ c \text{ (advertising outlays)}$$

The manner in which the period of observation is selected for estimating the values of a, b, and c under different circumstances is outlined below.

The Situation	*Period of Observation*
1. There is no basis for believing that any change has occurred in the relationship.	1923–41, 1947–53
2. The relationship may have changed in the postwar years.	a. Fit the function to the 1923–41 data; test the accuracy of its forecasts on 1947–53.
	(1) If tests show no appreciable loss in accuracy (based on statistical and other considerations), recompute estimates of a, b, and c using all the data.
	(2) If 1947–53 predictions are very poor in relation to estimates for 1923–41, and relationship in postwar years is evidently unstable, discard the function;
	or
	b. If relationship in postwar years seems stable, fit the function to 1947–53 data and test accuracy on 1923–41 data.

(1) If no appreciable loss in accuracy results, proceed as in (a, 1) above.

(2) If 1923–41 predictions are poor in relation to 1947–53 estimates, stick to the same (1947–53 fitted) function.

Note: The second series of alternatives is not very desirable here, since judging the stability of a relationship on the basis of seven observations is a risky affair.

The following additional comments might be offered with regard to selection of a period of observation.

1. Once the adequacy of a function for forecasting is established, maximum reliability is obtained by recomputing the coefficients as each new observation becomes available, provided that a change in the nature of the relationship is not evident. Thus, in the above example, once the 1954 forecast has been made and 1954 data becomes available, the estimates of the function's parameters for the 1955 forecast would be recomputed, based on the extended period of observation, 1923–41, 1947–54.

2. Other things being equal, the use of a longer period of observation is likely to produce more accurate predictions than a short period. At least this was the experience with aggregate consumption functions.

3. The practice sometimes advocated of selecting only those years for the period of observation that are felt to be more similar to the year being forecasted—such as using 1923–30, 1935–40 data to forecast 1948 conditions because the chances of 1948 being a depression year are small—is a dangerous one. Essentially, such a procedure begs the question, for to omit so-called atypical years is to assume implicitly that such conditions will not prevail in future years.[4] Furthermore, the applicability of functions derived in this manner to evaluating the relative importance of the various factors on sales is also restricted. Insofar as cyclical considerations enter into selection of a period of observation there is something to be said for attempting to include

[4] This criticism loses much of its force if the forecast function contains no lagged variable, for then the stage of business fluctuation in the forecasted period is likely to be determined implicitly by the values assigned to the independent variables.

experience over at least one complete cycle, with substantial differences between peaks and troughs.

Units of measurement

Judging again on the basis of tests carried out on a series of aggregate consumption functions, sales predictions based on regression functions are likely to be more accurate when the variables are adjusted for price changes and expressed in per capita units than when they are not. This does not take into consideration, however, possible errors resulting from faulty forecasts of price and population movements if a forecast is desired in aggregate terms at current prices.

Tests of predictive accuracy applied to a number of consumption functions, the variables of which were alternately adjusted for price changes, for population changes, for both, and for neither, indicated generally better accuracy when the price and population adjustments were made. These improvements were most pronounced for forecasts of level, the customary method, but were also evident for forecasts of change—using the first-difference forms of the function. It might be mentioned that little difference in predictive accuracy was observed as between functions whose variables were expressed in per capita units and functions in which population was used as a separate variable.

Choice of variables

We do not propose to recommend which (independent) variables should appear in a sales forecasting function: that is clearly a matter of individual circumstances. Rather we propose to put forth a few considerations that may be helpful in governing the selection of variables. Specifically, there are five.

AN ADDITIONAL CRITERION OF SELECTION

The principal criterion for selecting independent variables in a regression analysis is that they be related to the dependent variable, the variable under study. Since the resultant independent variables usually refer to the same time period as the dependent variable, a forecast of the latter cannot be obtained until the values of the independent variables are themselves forecasted. This is a fundamental weakness of the correlation technique when applied to forecasting, particularly in view of the relative neglect given to the matter of forecasting the independent variables. Thus, a prediction based on a function such as:

$$\text{sales} = a + b \ (\text{income})$$

may be in error not only because the relationship in the following year may differ slightly from what it was in the past, but also because the estimate of next year's income substituted in the equation is considerably in error. In many instances, the error in the sales forecast may be due more to the latter factor than to departure of the true relationship from the estimated one.

What is the answer? Clearly, effort must be made to reduce the error in the forecasts of independent variables as much as possible. Use of lagged variables is the ideal solution, but like most ideals this is rarely attainable. In its absence, one can utilize another criterion, which consists of giving preference to those independent variables that are most easily forecasted themselves. This criterion is meant to complement, rather than substitute for, the basic one cited above. Where a choice of independent variables is possible, its use may contribute substantially to reducing forecast errors. It is perhaps needless to add that, in any event, the forecast of the values of the independent variables deserves fully as much care and thought as is given to the derivation of the forecast function itself.

DEPENDENCE ON METHOD OF ESTIMATION

As a general rule, substantial differences in the accuracy of forecasts are obtained by changing the variables used in the regression function. Thus, the predictive accuracy of: sales $= a + b$ (income) may differ considerably from that of: sales $= a + b$ (income) $+ c$ (advertising). Surprising as it may seem, however, this is not the case when the functions are transformed into their first-difference forms and the forecasts made in terms of change. Once the most relevant variable has been taken into account, use of additional variables will probably make very little difference in predictive accuracy.

The reason for this derives from the manner in which the

forecasts are prepared by this method. As pointed out earlier, this involves adding the predicted change to last year's sales. But the latter is common to all functions. Hence, with this common benchmark and with the main determining variable already taken into account (by assumption), further refinements are likely to produce only slight changes in the final forecast.

Does this mean, then, that for all practical purposes it does not matter what variables are used (once the main determining variable is included)? No, unless one is completely sold on the superiority of this method of estimation. To ignore other relevant factors—especially possible changes in the composition of the aggregates used as the variables in the regression equation —is to ignore the possibility that a more "complete" function, in the sense of including a larger number of relevant factors, may yield higher predictive accuracy by the customary method of forecasting levels (as well as a better idea of the relative importance of various factors). The advantage of knowing that the first-difference method generally yields much the same accuracy for various functions is that, in a particular situation, it will probably suffice to test the superiority of this method by applying it to one function rather than to a number of them.

ADJUSTMENT FOR CYCLICAL EFFECTS

One of the most difficult problems in business and economic forecasting by regression methods is to secure a relationship that will remain invariant over the cycle. The general experience is to have a function approximate well the cyclical fluctuations in the sales variable during the period of observation, but then largely fail to trace the cyclical pattern of sales during the period being forecasted. In other words, the relationship tends to break down when used for prediction purposes.

A general solution to this problem still seems to be a long way off. It can be said, however, that a breakdown in a relationship seems most likely to occur when only variables relating to the same unit of time are employed, e.g., when current sales are related to current income and current advertising expenditures. In such a case, addition of a lagged variable that can be considered to reflect past influences on current sales may improve considerably the accuracy of the predictions. This is based on two findings with regard to aggregate savings-income relationships, namely:

1. Predictive accuracy of functions relating current savings to current income or current savings to current income and a time trend was improved substantially at times when a lagged income variable was added.

2. The only functions whose predictive accuracy was as good in the postwar years, 1947–52, when fitted to 1923–40 data as when fitted to 1923–30, 1935–40 data were those incorporating a lagged income variable. Otherwise, the predictive accuracy of functions fitted to the latter period of observation, which excludes the main depression years, was considerably better. Thus, lagged income in this case seems to have effectively counteracted the influence of the cycle.

Whether lagged income will be as effective in other situations is problematical. The important thing is to secure a variable that may be supposed to react with a lag on the variable being forecasted. In this case, it was lagged income; in another situation, it may be some other factor.

It might be noted that once the factor—here, income—has been selected, different forms of it may be used. In the previously mentioned study, two forms were used, both about equally effective. One form was income lagged one time unit (year), and the other form was the past cyclical peak value of income. The rationale behind the use of the latter is, in brief, that once people attain a certain level of living, their future expenditure and savings patterns when their income declines tend to be influenced by standards established earlier, reflecting in part a desire to regain their highest past level of living.

Needless to say, the use of more than one lagged variable is not undesirable. In fact, the ideal in a forecasting function would be to have all independent variables lagged, thus obviating the need for "guesstimating" the next period's values for these variables.

USE OF SERIAL CORRELATION

Marketing and economic data are known to be serially correlated. In other words, a company's sales one year are not independent of its previous year's sales. Next year's sales will be somewhere near current levels, almost surely within 15 percent. Sales are not going to double one year and then drop off 50 percent the next year (except perhaps for small firms).

This phenomenon of serial correlation has long been recognized and has been cited as a major limitation of the currently available means of assessing the adequacy of regression relationships. A close relationship between sales and income may at times be due as much to the serial correlation within each series as to the causal effect of one upon the other.

One means of getting around this problem has been the use of the first differences of the variables instead of the variables themselves. Thus one might correlate the change in sales with the change in income and with the change in advertising expenditures.

Another possibility, however, one that seems to have received relatively little attention, is to incorporate this serial correlation directly into the relationship. Try using last period's sales as an extra independent variable, i.e.:

current sales $= a + b$ (current income)
$+ c$ (current advertising expenditures)
$+ d$ (last period's sales)

Because of the serial correlation factor, the use of last period's sales in this manner can serve as a medium that reflects the effect of a multitude of past influences on the future level of sales, each of which influence alone may be too small to include separately or may not even be measurable.

Investigation of the autoregressive properties of sales (the serial correlation within the series) is especially desirable when a forecasting function is sought that is couched entirely in terms of lagged variables. In such cases, a function of the type [5]:

current sales $= a + b$ (last period's sales)
$+ c$ (change in sales from the preceding period
to the last period)

[5] The fact that all the illustrative equations cited in this article are linear—the simplest form—should not be interpreted as denying the value of more complex, nonlinear forms. Numerous examples of such functions fitted to consumption and employment data are to be found in the National Resources Committee publication, *Patterns of Resources Use* (U.S. Government Printing Office, Washington, D. C., 1938).

may yield highly accurate forecasts if sales are closely serially correlated and even identify turning points.

As a general rule, the value of serial correlation increases as the time unit decreases. A somewhat more complicated function was found by the writer to predict railroad shipper's expected quarterly carloadings remarkably well, purely by utilizing the serial correlation properties of actual carloadings. In another study, it was found that quarterly consumption expenditures were related about as closely to the preceding quarter's expenditures as to any other lagged variable, including income.

VALUE OF A TIME TREND

It is customary to insert a time variable in many regression functions to reflect changes in tastes, technological trends, and other miscellaneous factors. The value of such a procedure for prediction purposes is open to question. The reason for this, basically, is that time by itself is not a cause of business or economic changes. Time is the *medium* through which events take place, but it is not the cause. Thus, one may broil a steak over a fire. The broiling takes a certain amount of time but the reason the steak gets broiled is not the passage of time but the presence of the fire. Without the fire, the broiling would not occur no matter how much time were to pass.

A statistically significant coefficient for a time variable in a forecast function shows, in effect, that certain relevant causal variables are omitted from the relationship and that the time trend is acting as a substitute for them. Use of such a relationship for prediction therefore involves the implicit assumption that the relations between time and these other variables will continue in the same manner as before. Without even knowing what these other variables are, as is usually the case, this is a dangerous assumption. Thus the United States population increased in a more or less linear arithmetic fashion during the 1930s and then rose more rapidly during the 1940s. Any function fitted to the 1930s and including time instead of population as an independent variable would have gone haywire as a result if used for prediction purposes in the 1940s.

The upshot of this discussion is that wherever time is found to be a significant variable in a sales forecasting function, a spurious relation is involved in the sense that one or more causal variables are being omitted. If identified, they should be substituted for the time variable. In many cases, they will not be found, they may not be measurable, or they may be too numerous to be included in the function. The alternative then is to retain the time variable, but recognition should be given to the danger presented from this source in making predictions with that function.[6]

EVALUATING THE ACCURACY OF FORECASTS

Evaluating and keeping a record of the accuracy of the forecasts is fully as important as making the forecasts, for it is only by some such means that clues can be obtained to improving the accuracy of future forecasts. This is true irrespective of the forecasting method employed, and yet it is a consideration that appears to be overlooked in more cases than not.[7]

There are many ways of measuring accuracy, and of those three will be advanced here. Before doing so, however, we begin with a note of caution on how not to evaluate the accuracy of forecasts. This may be a rather surprising one: Don't rely on the coefficient of correlation.

The coefficient of correlation is the generally accepted measure of the goodness of fit of a function to the observations. It has therefore been more or less implicitly assumed that the same statistic would also indicate the adequacy of a function for pre-

[6] A new use of a time variable, proposed by Ashley Wright, is using it as an independent check on sales forecasts by predicting sales from a linear relation between sales and time. The advantage of this procedure is that there is then no problem of predicting the value of the independent variable, time. A forecast based on such a relation is, of course, nothing more than an extrapolation of trend. However, in the case of products that possess reasonably stable trends, such a forecast is a valuable check on the forecasts derived by other means because, in such cases, year-to-year changes in sales are rarely likely to deviate substantially from the projected trend value. Hence, a company forecast which is, say, four standard errors away from the projected trend forecast would be immediately subject to question.
[7] A better approximation to the standard error of the forecast is often obtainable through historical experience than by straight substitution in statistical formulas.

diction purposes. For if a function provides a good fit to the observed data, is it not logical to assume that it would yield better forecasts than one that provides a poor fit to the observed data?

The answer is that it may be logical but, as in the case of Porgy and Bess, "it ain't necessarily so." In fact, in the case of the writer's work with aggregate savings functions, no relationship at all was found between the coefficient of correlation of a function for the period of observation and the accuracy of the function's predictions.[8] Where different periods of observation were involved, the relationship between the coefficient of correlation and predictive accuracy was, if anything, negative— meaning that for the functions studied, one with a poor goodness of fit to the observed data tended to yield more accurate forecasts than one with a better goodness of fit!

Does this result then lead to the fantastic implication that functions whose forecasts are most accurate are likely to be the ones with the poorest coefficients of correlation? No, for it can be shown that special factors may be present with different periods of observation which completely distort the value of the coefficient of correlation as an indicator of predictive accuracy.[9] Even with the same period of observation and with the same variable being estimated, however, it does seem that little reliance can be placed on the coefficient of correlation as an a priori measure of predictive accuracy. The same is also true of another statistic relating to the period of observation that was tested, the average absolute relative margin of error.

The reason for the poor showing of these measures may well be the tendency to place too much emphasis on inserting those variables in a function that will yield a good fit in a particular situation and too little consideration to the more basic factors that influence the variable over a long period of time. This is

[8] The same absence of relationship was found for functions explaining railroad shippers' forecasts and later used to make the forecasts.

[9] The explanation is that the shorter period in the savings study, 1929–40, increased the relative importance of the depression years, 1931–34. This raised the coefficient of correlation because of the resultant greater amplitude of the dependent variable and at the same time reduce the accuracy of the predictions of the postwar (prosperous) years. On the other hand, the use of the longer period of observation, 1923–40, by reducing the importance of the depression years, tended to lower the coefficient of correlation but also raised the postwar predictive accuracy of the functions.

admittedly only a hypothesis, and much further work needs to be done to determine its validity and give it operational meaning.

We close this discussion of the coefficient of correlation with what is probably another unsettling comment regarding its value, namely:

A comparison of the coefficients of correlations of two functions based on two different periods of observation or involving two different dependent variables provides no indication of the relative accuracy of the functions *even in the period of observation.*

In other words, the relative magnitudes of the coefficients of correlation may yield no information regarding the size of the residuals or the relative errors of the function estimates in the period of observation.

This point is illustrated in the chart, which is reproduced in part from the author's savings study, and which shows the coefficients of correlation, the residuals, and the periods of observation for six different forms of aggregate savings functions. The chart shows in rather striking fashion how (a) the same function fitted to different periods of observation can have widely different coefficients of correlation although the residuals for overlapping years in the period of observation are almost identical, and (b) the accuracy of the predictions of two forms of the same function, each fitted to a different period of observation, differs markedly with, if anything, negative relationship to the coefficient of correlation.

The explanation of both phenomena lies in the differences in amplitude during the period of observation of the variable being estimated—savings, in this case. When the amplitude of the dependent variables of two functions differ—whether because of differences in the period of observation, in the units of measurement, in the variables used, or because of differences in definitions—no reliance can be placed on a comparison of the coefficients of correlation from the point of view of either general adequacy of the functions or their predictive accuracy.

Three standards of accuracy

There are at least three ways in which the accuracy of a set of predictions may be readily evaluated, each

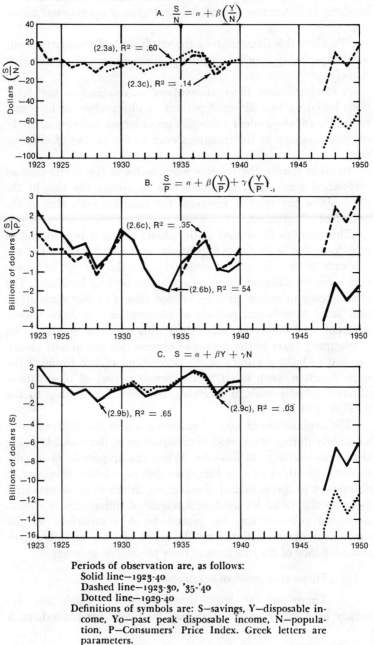

CHART 1. RESIDUALS OF SELECTED SAVINGS FUNCTIONS

A. $\frac{S}{N} = \alpha + \beta \left(\frac{Y}{N}\right)$

B. $\frac{S}{P} = \alpha + \beta \left(\frac{Y}{P}\right) + \gamma \left(\frac{Y}{P}\right)_{-1}$

C. $S = \alpha + \beta Y + \gamma N$

Periods of observation are, as follows:
Solid line—1923-40
Dashed line—1923-30, '35-'40
Dotted line—1929-40
Definitions of symbols are: S—savings, Y—disposable income, Yo—past peak disposable income, N—population, P—Consumers' Price Index. Greek letters are parameters.

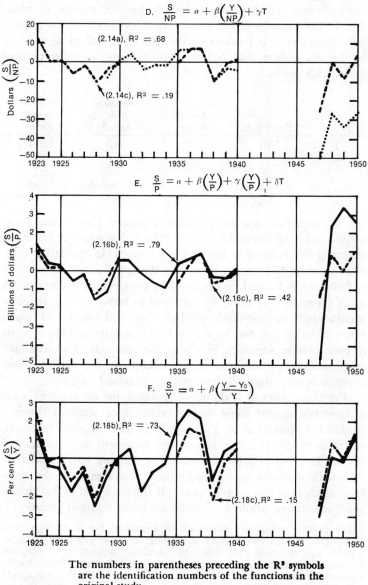

D. $\dfrac{S}{NP} = \alpha + \beta\left(\dfrac{Y}{NP}\right) + \gamma T$

(2.14a), $R^2 = .68$

(2.14c), $R^2 = .19$

E. $\dfrac{S}{P} = \alpha + \beta\left(\dfrac{Y}{P}\right) + \gamma\left(\dfrac{Y}{P}\right)_1 + \delta T$

(2.16b), $R^2 = .79$

(2.16c), $R^2 = .42$

F. $\dfrac{S}{Y} = \alpha + \beta\left(\dfrac{Y - Y_0}{Y}\right)$

(2.18b), $R^2 = .73$

(2.18c), $R^2 = .15$

The numbers in parentheses preceding the R^2 symbols are the identification numbers of the functions in the original study.

Source: Reproduced by permission of the National Bureau of Economic Research from *A Study of Aggregate Consumption Functions*, pp. 52-53.

possessing distinctive advantages and disadvantages. Perhaps the most obvious is the average absolute percent margin of error, that is, the average of the percentage deviations of the predictions from the actual values taken without regard to sign. Thus, in the illustration that follows, the average absolute percent error is 8.8 percent.[10]

Period	Prediction	Actual Value	Deviation from Actual	Absolute Percent Error
1	70	80	−10	12.5%
2	105	100	5	5.0
3	98	90	8	8.9
			Average =	8.8%

Although this measure provides a general idea of the accuracy of a set of forecasts, it is of little value when it comes to assessing the relative value of the forecasts. In the latter case, what we want to know is: How does this set of forecasts compare to forecasts that could be prepared by some simple technique? A set of forecasts may show up very well in terms of their absolute accuracy, as measured above, but they will not be of much practical value if it can be shown that a simpler technique leads to even greater accuracy. Similarly, the particular forecasts may err substantially, but if no other method can be made to yield superior results, they may still possess practical value.

For this reason, it is desirable to assess the accuracy of a set of forecasts against some simple yardstick of accuracy. Such a yardstick is provided by the so-called "naive model" forecasts, which postulate that the future will represent an extension of present levels or of present trends. The simplest naive model for annual data would be to predict next year's sales at the same level as those of the current year. (If the time unit is less than a year, adjustment would have to be made for seasonal variation.)

[10] The practice sometimes employed of averaging the percentage error with regard to sign can be very misleading, as it permits large errors of opposite sign to cancel each other. In the above illustration, for example, the procedure would produce an average error of 1.4 percent, which hardly represents the true accuracy. This does not mean, however, that one should ignore the signs of the deviations in other respects, for it is often through study of the signs that bias in forecasts is detected.

Such a forecast is nothing more than an extension of present levels. For firms and industries still in growth stages, a naive-model projection of present trends would be more realistic, such as predicting that:

$$\text{next year's sales} = \text{this year's sales} \times \frac{\text{this year's sales}}{\text{last year's sales}}$$

Alternatively, one could predict that the increase in next year's sales would be only one-half of the increase over the past year, or any other fraction that seems plausible in the particular case.

By comparing the predictive accuracy of the regression functions or methods under consideration with that of one or more of these naive models, a good idea is obtained of the practical value of these more elaborate methods from the company's viewpoint, and whether the latter are worth using. In this connection, it might be noted that a result in favor of a naive model does not provide a basis for using it on a continuing basis (unless one has evidence that this particular model is not so "naive" as it was made out to be). Rather, it points to the need for further exploration to derive forecasting methods that will yield better results on a sound foundation.

The third standard of accuracy is one designed to remedy a fault in the first one. This is due to the fact that any comparison of levels of forecasts with actual sales is almost bound to appear in a favorable light because of the serial correlation in sales. Sales do not change much from one year (or period, after adjustment for seasonal variation) to the next—especially as the level of aggregation rises. Not only is next year's sales related to this year's level, but the prediction for the next year is also invariably closely related to this year's level. Hence a spurious relationship is introduced between the prediction and the actual figure which, in any comparison of levels, tends to produce an unwarrantedly high impression of accuracy.

A simple means of removing this spurious factor (at least, in many cases) is to compare the directions of change rather than the levels. In this manner, any association between the forecasts and actual figures due to level is largely removed. This procedure also has the advantage of indicating the success of the forecasting method in gauging the direction of movement in sales, one of the central problems in business forecasting.

The effect of changing the standard of comparison from one of level to one of direction of change was brought out vividly in a recent study evaluating the accuracy of a set of forecasts of quarterly railroad carloadings. Over a period of about 25 years, these carloadings on an aggregate basis came to within 9 percent of the true figure on the average. At the same time, however, no relationship at all was found between the anticipated *change* in the shipments and the actual change. Surprising as this result may seem, it may well be typical of business and economic forecasts: they yield good approximations of the general level of business but are as likely as not to miss the direction of change.

Finally a few words would seem useful on the desirability of keeping records of the variability of the accuracy produced by various forecasting functions or methods. If two functions yield similar average errors, the one with the lower variability is to be preferred, other considerations being equal, for it is that method in which greater confidence can be placed for planning purposes.

GENERAL CONSIDERATIONS

Although this article has dealt with the use of correlation techniques in sales forecasting, it does not mean to advocate the general use of this technique. Correlation and regression functions is only one of a number of methods available for sale forecasting, and the method to use in a particular case depends on the circumstances involved. Often, it is desirable to employ more than one method in order to check a forecast by more or less independent means. Where correlation techniques are used, however, this article has attempted to draw attention to various means that might be employed to improve their efficiency.

What Do Statistical "Demand Curves" Show?

E. J. WORKING

The late E. J. Working was Professor of Economics at the
University of Illinois. This classic article first appeared in
the *Quarterly Journal of Economics* in 1927.

Let us first consider in what way statistical demand
curves are constructed. While both the nature of the data used
and the technique of analysis vary, the basic data consist of corresponding prices and quantities. That is, if a given quantity refers
to the amount of commodity sold, produced, or consumed in the
year 1910, the corresponding price is the price that is taken to be
typical for the year 1910. These corresponding quantities and
prices may be for a period of a month, a year, or any other
length of time that is feasible; and, as has already been indicated,
the quantities may refer to amounts produced, sold, or consumed. The technique of analysis consists of such operations as
fitting the demand curve and adjusting the original data to remove, in so far as is possible, the effect of disturbing influences.
For a preliminary understanding of the way in which curves are
constructed, we need not be concerned with the differences in
technique; but whether the quantities used are the amounts

produced, sold, or consumed is a matter of greater significance, which must be kept in mind.

For the present, let us confine our attention to the type of study that uses for its data the quantities which have been sold in the market. In general, the method of constructing demand curves of this sort is to take corresponding prices and quantities, plot them, and draw a curve that will fit as nearly as possible all the plotted points. Suppose, for example, we wish to determine the demand curve for beef. First, we find out how many pounds of beef were sold in a given month and what was the average price. We do the same for all the other months of the period over which our study is to extend, and plot our data with quantities as abscissas and corresponding prices as ordinates. Next we draw a curve to fit the points. This is our demand curve.

In the actual construction of demand curves, certain refinements necessary in order to get satisfactory results are introduced. The purpose of these is to correct the data so as to remove the effect of various extraneous and complicating factors. For example, adjustments are usually made for changes in the purchasing power of money, and for changes in population and in consumption habits. Corrections may be made directly by such means as dividing all original price data by "an index of the general level of prices." They may be made indirectly by correction for trends of the two time series of prices and of quantities. Whatever the corrections and refinements, however, the essence of the method is that certain prices are taken as representing the prices at which certain quantities of the product in question were sold.

With this in mind, we may now turn to the theory of the demand-and-supply curve analysis of market prices. The conventional theory runs in terms substantially as follows. At any given time all individuals within the scope of the market may be considered as being within two groups—potential buyers and potential sellers. The higher the price, the more the sellers will be ready to sell and the less the buyers will be willing to take. We may assume a demand schedule of the potential buyers and a supply schedule of the potential sellers which express the amounts that these groups are ready to buy and sell at different prices.

From these schedules supply and demand curves may be made. Thus we have our supply and demand curves showing the market situation at any given time, and the price that results from this situation will be represented by the height of the point where the curves intersect.

This, however, represents the situation as it obtains at any given moment only. It may change; indeed, it is almost certain to change. The supply and demand curves which accurately represent the market situation of to-day will not represent that of a week hence. The curves that represent the average or aggregate of conditions this month will not hold true for the corresponding month of next year. In the case of the wheat market, for example, the effect of news that wheat which is growing in Kansas has been damaged by rust will cause a shift in both demand and supply schedules of the traders in the grain markets. The same amount of wheat, or a greater, will command a higher price than would have been the case if the news had failed to reach the traders. Since much of the buying and selling is speculative, changes in the market price itself may result in shifts of the demand and supply schedules.

If, then, our market demand and supply curves are to indicate conditions that extend over a period of time, we must represent them as shifting. A diagram such as the following, Fig. 1, may be used to indicate them. The demand and supply curves may meet at any point within the area a, b, c, d, and over a period of time points of equilibrium will occur at many different places within it.

But what of statistical demand curves in the light of this analysis? If we construct a statistical demand curve from data of quantities sold and corresponding prices, our original data consist, in effect, of observations of points at which the demand and supply curves have met. Although we may wish to reduce our data to static conditions, we must remember that they originate in the market itself. The market is dynamic and our data extend over a period of time; consequently our data are of changing conditions and must be considered as the result of shifting demand and supply schedules.

Let us assume that conditions are such as those illustrated in Fig. 2, the demand curve shifting from D_1 to D_2, and the supply

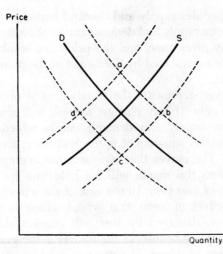

Price

D   S

a

d   b

c

Quantity

FIG. 1

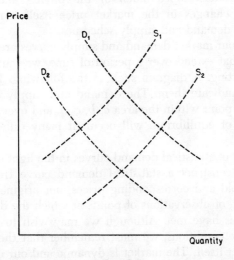

Price

D_1 S_1

D_2 S_2

Quantity

FIG. 2

curve shifting in similar manner from S_1 to S_2. It is to be noted that the diagram shows approximately equal shifting of the demand and supply curves.

Under such conditions there will result a series of prices which may be graphically represented by Fig. 3. It is from data such as those represented by the dots that we are to construct a demand curve, but evidently no satisfactory fit can be obtained. A line of one slope will give substantially as good a fit as will a line of any other slope.

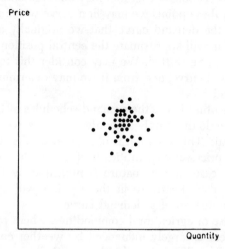

Fig. 3

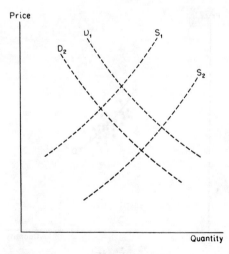

Fig. 4

But what happens if we alter our assumptions as to the relative shifting of the demand and supply curves? Suppose the supply curve shifts in some such manner as is indicated by Fig. 4, that is, so that the shifting of the supply curve is greater than the shifting of the demand curve. We shall then obtain a very different set of observations—a set that may be represented by the dots of Fig. 5. To these points we may fit a curve which will have the elasticity of the demand curve that we originally assumed, and whose position will approximate the central position about which the demand curve shifted. We may consider this to be a sort of typical demand curve, and from it we may determine the elasticity of demand.

If, on the other hand, the demand schedules of buyers fluctuate more than do the supply schedules of sellers, we shall obtain a different result. This situation is illustrated by Fig. 6. The resulting array of prices and quantities is of a very different sort from the previous case, and its nature is indicated by Fig. 7. A line drawn so as most nearly to fit these points will approximate a supply curve instead of a demand curve.

In the case of agricultural commodities, where production for any given year is largely influenced by weather conditions, and where farmers sell practically their entire crop regardless of

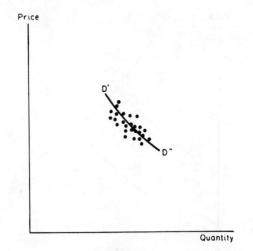

Fɪɢ. 5

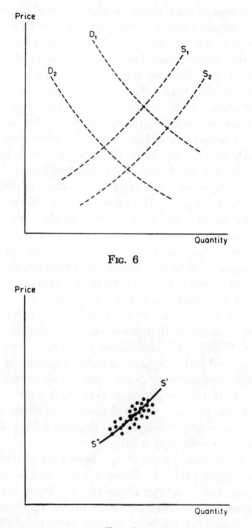

FIG. 6

FIG. 7

price, there is likely to be a much greater shifting of the supply schedules of sellers than of the demand schedules of buyers. This is particularly true of perishable commodities, which cannot be withheld from the market without spoilage, and in case the farmers themselves can under no conditions use more than a very small proportion of their entire production. Such a condition results in the supply curve's shifting within very wide limits. The demand curve, on the other hand, may shift but little. The quantities that are consumed may be dependent almost entirely upon price, so that the only way to have a much larger amount taken off the market is to reduce the price, and any considerable curtailment of supply is sure to result in a higher price.

With other commodities, the situation may be entirely different. Where a manufacturer has complete control over the supply of the article that he produces, the price at which he sells may be quite definitely fixed, and the amount of his production will vary, depending upon how large an amount of the article is bought at the fixed price. The extent to which there is a similar tendency to adjust sales to the shifts of demand varies with different commodities, depending upon how large overhead costs are and upon the extent to which trade agreements or other means are used to limit competition between different manufacturers. In general, however, there is a marked tendency for the prices of manufactured articles to conform to their expenses of production, the amount of the articles sold varying with the intensity of demand at that price which equals the expenses of production. Under such conditions, the supply curve does not shift greatly, but rather approximates an expenses-of-production curve, which does not vary much from month to month or from year to year. If this condition is combined with a fluctuating demand for the product, we shall have a situation such as that shown in Figs. 6 and 7, where the demand curves shift widely and the supply curves only a little.

From this, it would seem that, whether we obtain a demand curve or a supply curve, by fitting a curve to a series of points that represent the quantities of an article sold at various prices, depends upon the fundamental nature of the supply and demand conditions. It implies the need of some term in addition to that of elasticity in order to describe the nature of supply and demand. The term "variability" may be used for this purpose. For example, the demand for an article may be said to be "elastic" if, at a given time, a small reduction in price would result in a much greater quantity's being sold, while it may be said to be "variable" if the demand curve shows a tendency to shift markedly. To be called variable, the demand curve should have the tendency to shift back and forth, and not merely to shift gradually and consistently to the right or left because of changes of population or consuming habits.

Whether a demand or a supply curve is obtained may also be affected by the nature of the corrections applied to the original data. The corrections may be such as to reduce the effect of the shifting of the demand schedules without reducing the effect of the shifting of the supply schedules. In such a case the curve obtained will approximate a demand curve, even though the original demand schedules fluctuated fully as much as did the supply schedules.

By intelligently applying proper refinements and making corrections to eliminate separately those factors that cause demand curves to shift and those factors that cause supply curves to shift, it may be possible even to obtain both a demand curve and a supply curve for the same product and from the same original data. Certainly it may be possible, in many cases where satisfactory demand curves have not been obtained, to find instead the supply curves of the articles in question. The supply curve obtained by such methods, it is to be noted, would be a market supply curve rather than a normal supply curve.

Thus far it has been assumed that the supply and demand curves shift quite independently and at random; but such need not be the case. It is altogether possible that a shift of the demand curve to the right may, as a rule, be accompanied by a shift of the supply curve to the left, and vice versa. Let us see what result is to be expected under such conditions. If successive positions of the demand curve are represented by the curves, D_1, D_2, D_3, D_4, and D_5 of Fig. 8, while the curves S_1, S_2, S_3, S_4, and S_5 represent corresponding positions of the supply curves, then a series of prices will result from the intersection of D_1 with S_1, D_2 with S_2, and so on. If a curve be fitted to these points, it will not conform to the theoretical demand curve. It will have a smaller elasticity, as is shown by $D'D''$ of Fig. 8. If, on the other hand, a shift of the demand curve to the right is accompanied by a shift of the supply curve to the right, we shall obtain a result such as that indicated by $D'D''$ in Fig. 9. The fitted curve again fails to

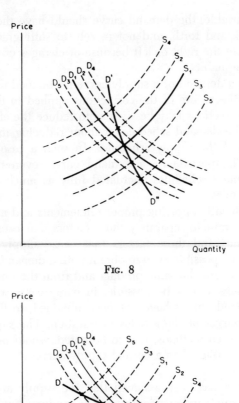

FIG. 8

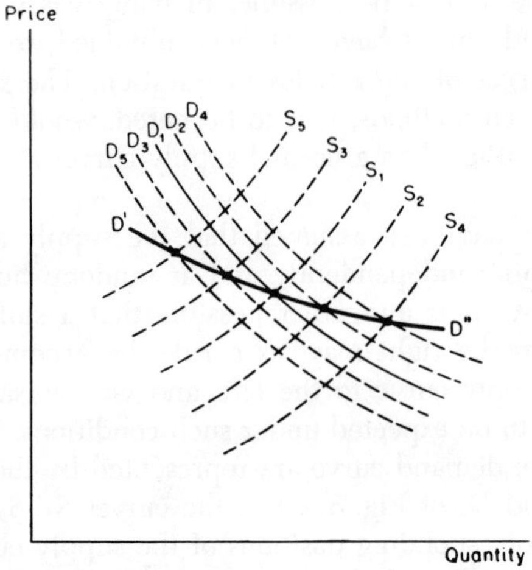

FIG. 9

conform to the theoretical one, but in this case it is more elastic.

Without carrying the illustrations further, it will be apparent that similar reasoning applies to the fitted "supply curve" if case conditions are such that the demand curve shifts more than does the supply curve.

If there is a change in the range through which the supply curve shifts, as might occur through the imposition of a tariff on an imported good, a new fitted curve will result, which will not be a continuation of the former one—this because the fitted curve does not correspond to the true demand curve. In case, then, of correlated shifts of the demand and supply curves, a fitted curve cannot be considered to be the demand curve for the article. It cannot be used, for example, to estimate what change in price would result from the levying of a tariff upon the commodity.

Perhaps a word of caution is needed here. It does not follow from the foregoing analysis that, when conditions are such that shifts of the supply and demand curves are correlated, an attempt to construct a demand curve will give a result that will be useless. Even though shifts of the supply and demand curves are correlated, a curve that is fitted to the points of intersection will be useful for purposes of price forecasting, provided no new factors are introduced which did not affect the price during the period of the study. Thus, as long as the shifts of the supply and demand curves remain correlated in the same way, and as long as they shift through approximately the same range, the curve of regression of price upon quantity can be used as a means of estimating price from quantity.

In cases where it is impossible to show that the shifts of the demand and supply curves are not correlated, much confusion would probably be avoided if the fitted curves were not called demand curves (or supply curves), but if, instead, they were called merely lines of regression. Such curves may be useful, but we must be extremely careful in our interpretation of them. We must also make every effort to discover whether the shifts of the supply and demand curves are correlated before interpreting the results of any fitted curve.

Statistical Cost Functions of a Hosiery Mill

JOEL DEAN

Joel Dean was formerly Professor of Business Economics
at Columbia University. This paper appeared as part of a
supplement to the *Journal of Business* in 1941.

The enterprise whose cost behavior was analyzed is a
hosiery knitting mill that is one of a number of subsidiary plants
of a large silk-hosiery manufacturing firm. In the particular plant
studied, the manufacturing process is confined to the knitting of
the stockings, that is, the plant begins with the wound silk and
carries the operations up to the point where the stockings are
ready to be shipped to other plants for dyeing and finishing.
The operations in the mill are, therefore, carried on by highly
mechanized equipment and skilled labor.

Cost functions were determined for combined cost and for
its components: productive labor cost, nonproductive labor cost,
and overhead cost. These functions were derived separately for
monthly, quarterly, and weekly data. For the monthly and
quarterly observations both simple and partial regressions of
the various costs on output were obtained. In this paper the
statistical findings for the monthly data alone are presented.

1. SIMPLE REGRESSIONS

Scatter diagrams were made between output and combined cost and its three components for the monthly data to
indicate the form of the restricted cost function in which out-

put is the only independent variable. The simple regression [1] indicated by the scatter diagrams appeared to be linear, so that a regression equation of the first degree with the general form $X_1 = b_1 + b_2X_2$ was fitted to the observations for combined cost and its three components.[2] The regression equations derived for the four categories of cost in the form of monthly totals, together with the statistical constants, are shown in Table 1.

TABLE 1*

HOSIERY MILL: SIMPLE REGRESSIONS OF COMBINED COST AND ITS COMPONENTS ON OUTPUT
(MONTHLY OBSERVATIONS)

	Combined Cost	Productive Labor Cost	Nonproductive Labor Cost	Overhead Cost
Simple regression equation	$cX_1 = 2935.59$ $+1.998X_2$	$pX_1 = -1695.16$ $+ 1.780X_2$	$nX_1 = 992.23$ $+ 0.097X_2$	$oX_1 = 3638.30$ $+ 0.121X_2$
Standard error of estimate	6109.83	5497.09	399.34	390.58
Correlation coefficient (r).	0.973	0.972	0.952	0.970
Regression coefficient (b).	1.998 ± 0.034	1.780 ± 0.035	0.097 ± 0.045	0.121 ± 0.036

* The symbols have the following meaning:

cX_1 = combined cost in dollars

pX_1 = productive labor cost in dollars

nX_1 = nonproductive labor cost in dollars

oX_1 = overhead cost in dollars

X_2 = output in dozens of pairs

The results which are expressed in a mathematical form in Table 1 can also be shown by regression lines or scatter diagrams. The regression equations for combined cost and productive labor cost are illustrated graphically in Chart 1. Chart 2 shows the simple regressions not only of the aggregate nonproductive labor cost, but also of its principal elements: supervision, maintenance, labor, office staff, and other indirect labor. In order to show more clearly the nature of the individual cost functions, each of the cost elements and their total are measured from a common base,

[1] The "simple" regression referred to should be carefully distinguished from the "net" or "partial" regressions.

[2] Furthermore, statistical examination of the relation of cost and output first differences (an approximation to marginal cost) and of the relation of average cost to output supported the hypothesis of linearity of the total cost function. Despite the support given the linear total cost specification by the analysis of the production techniques, by the distribution of total cost observations, and by the behavior of average cost and the approximation to marginal cost, a cubic function was also specified and fitted by least-squares regression analysis. The higher-order function did not appear to fit the data significantly better than the linear function.

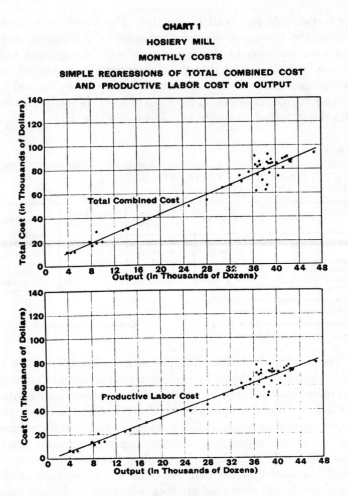

CHART 1

HOSIERY MILL

MONTHLY COSTS

SIMPLE REGRESSIONS OF TOTAL COMBINED COST AND PRODUCTIVE LABOR COST ON OUTPUT

the X axis, that is, they are not cumulated. Simple regressions of total overhead cost and its elements are similarly presented in Chart 3.

2. PARTIAL REGRESSIONS*

Graphic multiple correlation analysis showed that the deviations from the simple regression functions of cost on out-

* Readers who are familiar only with simple regression can skip this section and go directly to section 3. Editor.

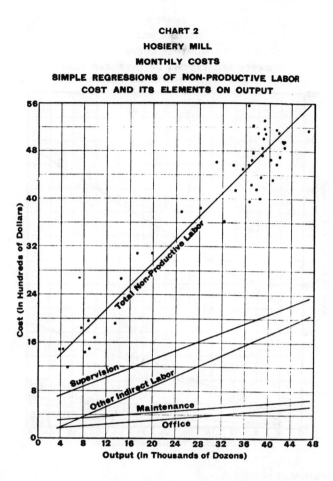

CHART 2

HOSIERY MILL

MONTHLY COSTS

**SIMPLE REGRESSIONS OF NON-PRODUCTIVE LABOR
COST AND ITS ELEMENTS ON OUTPUT**

put were systematically ordered in time. This indicated that a correction for a time trend might be advisable. A time factor was, therefore, introduced explicitly into the least-squares multiple correlation analysis by the use of the variable X_3, which is a series consisting of the sequential numbering of the months in which observations were taken. In this way it was possible to isolate the systematic variation of cost as a function of time and to determine the net regression of cost on output. By allowing for the influences of changes in conditions through time, which had not been taken into account by the rectification of the data, an estimate of the cost-output function which was possibly more

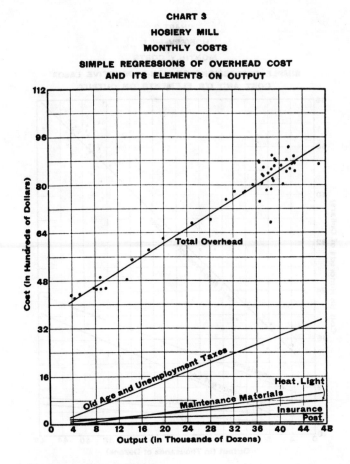

CHART 3

HOSIERY MILL

MONTHLY COSTS

SIMPLE REGRESSIONS OF OVERHEAD COST
AND ITS ELEMENTS ON OUTPUT

accurate was obtained.

The graphic analysis showed a significant time trend for the three major cost components—productive labor, nonproductive labor and overhead—as well as for combined cost. The graphic partial regression of cost and time appeared to be curvilinear in the case of combined cost, productive labor cost, and overhead cost. A curvilinear multiple regression equation of the general form,

$$X_1 = b_1 + b_2 X_2 + b_3 X_3 + b_4 X_3^2$$

was therefore selected as the most appropriate specification in these cases. This equation retains the linear specification chosen in the case of the simple regression, since this multiple regression equation is still linear with respect to output. In the remaining instance—nonproductive labor—a linear function, $X_1 = b_1 + b_2 X_2 + b_3 X_3$, was fitted. In these equations X_1 is cost (in the form of totals per month), X_2 is output (in dozens of pairs), and X_3 is time (months numbered sequentially).

The results of the multiple correlation analysis of the monthly data for combined cost and its three principal components are shown mathematically in Table 2 on page 212. These findings are also displayed in graphic form in the accompanying charts (4, 5, 6, and 7), in which the net or partial regressions of the various cost categories on output and time are shown.

In the upper section of Chart 4 the dots represent rectified monthly totals of combined cost that have been adjusted for the curvilinear time trend shown in the lower sections of the chart. Although the scatter is considerable, the distribution of the dots appears to substantiate the linearity of the partial regression of total cost over the observed range, from 4000 dozen to 43,000 dozen pairs of hosiery. Beyond this range there is only one observation. The irregular line in the lower section connects cost observations that have been adjusted for output. They are deviations of the observations from the simple regression of cost on output arranged chronologically. The curved line fitted to these ordered deviations is the partial regression of cost on time, which is assigned a magnitude by the sequential numbering of the months.

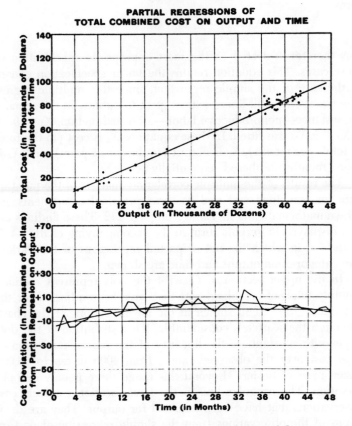

CHART 4

HOSIERY MILL

MONTHLY COSTS

PARTIAL REGRESSIONS OF
TOTAL COMBINED COST ON OUTPUT AND TIME

A parallel portrayal of variations in productive labor cost with respect to output and time is found in Chart 5. The distribution of adjusted observations of monthly costs plotted against output in the upper section appears to be linear. As in the preceding chart, cost observations have been adjusted for the curvilinear partial regression of cost deviations on time shown in the lower section of the chart.

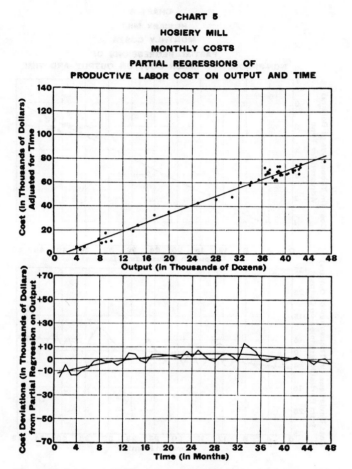

CHART 5

HOSIERY MILL

MONTHLY COSTS

PARTIAL REGRESSIONS OF

PRODUCTIVE LABOR COST ON OUTPUT AND TIME

Chart 6 shows partial regressions for monthly totals of non-productive labor cost. In the upper section are plotted corrected cost observations adjusted for time trend. Again the amount of scatter and the character of the distribution of dots does not appear to justify specification of other than a linear partial regression. The deviations from this output regression, which were arranged chronologically and connected by an irregular line in the lower section of the chart, indicate a steady upward trend in nonproductive labor cost after allowance is made for the effect of output.

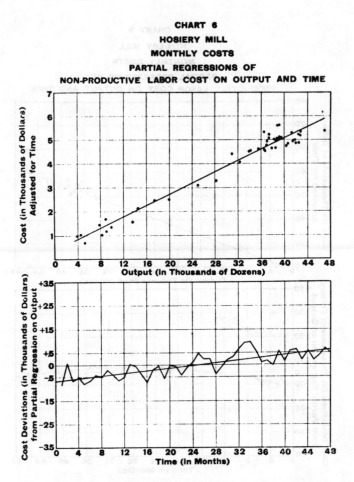

CHART 6

HOSIERY MILL

MONTHLY COSTS

PARTIAL REGRESSIONS OF

NON-PRODUCTIVE LABOR COST ON OUTPUT AND TIME

Chart 7 shows the partial regressions and adjusted observations of total monthly overhead cost. The scatter of adjusted cost observations plotted against output (shown by the dots in the upper section) is so wide and so approximately linear that fitting a cubic or parabolic regression curve does not appear to be justified. The linear partial regression shows that total overhead cost tends to increase with output at a uniform rate over the volume range studied. The lower section shows the time trend in overhead cost behavior, when allowance is made for the effects of output. The irregular line shows chronologically ordered cost deviations from the regression line appearing in the upper section of the chart. The trend is indicated by the curvilinear partial regression. There appeared to be a general tendency for overhead cost to increase during the first part of the period, to level off, and then to decline somewhat in the later months.

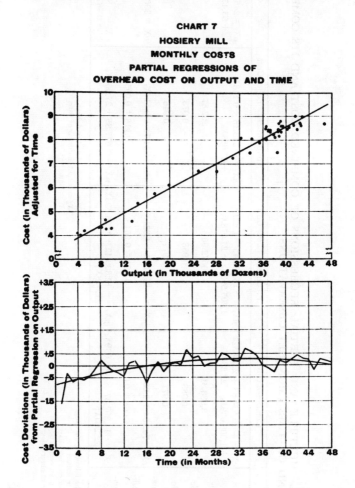

CHART 7

HOSIERY MILL

MONTHLY COSTS

PARTIAL REGRESSIONS OF
OVERHEAD COST ON OUTPUT AND TIME

TABLE 2*
HOSIERY MILL: MULTIPLE AND PARTIAL REGRESSIONS OF COMBINED COST AND ITS COMPONENTS ON OUTPUT AND ON TIME (MONTHLY OBSERVATIONS)

	Combined Cost	Productive Labor Cost	Nonproductive Labor Cost	Overhead Cost
Multiple regression equation	$cX_1 = -13,634.83 + 2.068X_2 + 1,308.039X_3 - 22.280X_3^2$	$pX_1 = -15,832.45 + 1.821X_2 + 1,205.593X_3 - 21.078X_3^2$	$nX_1 = -343.15 + 0.118X_2 + 27.668X_3$	$oX_1 = 2451.60 + 0.130X_2 + 65.457X_3 - 0.987X_3^2$
Standard error of estimate †	3,983.31	3572.90	302.87	296.57
Coefficient of multiple correlation $\bar{R}$ †	0.988	0.988	0.973	0.983
Coefficient of multiple determination $\bar{R}^2$ †	0.977	0.977	0.946	0.966
Partial regression equation for output	$cX_1 = 762.54 + 2.068X_2$	$pX_1 = -2993.03 + 1.821X_2$	$nX_1 = 334.71 + 0.118X_2$	$oX_1 = 3363.47 + 0.130X_2$
Partial regression coefficient for output	2.068 ± 0.071	1.821 ± 0.064	0.118 ± 0.005	0.130 ± 0.005
Partial regression equation for time	$cX_1 = -14,397.37 + 1308.039X_3 - 22.280X_3^2$	$pX_1 = -12,839.42 + 1205.593X_3 - 21.078X_3^2$	$nX_1 = -677.85 + 27.668X_3$	$oX_1 = -821.87 + 65.458X_3 - 0.987X_3^2$

* The symbols have the following meaning:
cX_1 = combined cost in dollars
pX_1 = productive labor cost in dollars
nX_1 = nonproductive labor cost in dollars
oX_1 = overhead cost in dollars
X_2 = output in dozens of pairs
X_3 = time
† Adjusted for number of observations.

3. MARGINAL AND AVERAGE COST

Both the simple and the partial regressions on output were determined for costs in the form of monthly totals. Both types of total cost functions were transformed [3] into average and marginal cost functions.[4] The equations for the derived average and marginal cost functions for combined cost and for its major components are found in Table 3.

TABLE 3*

HOSIERY MILL: EQUATIONS FOR TOTAL, AVERAGE AND MARGINAL COST-OUTPUT FUNCTIONS OBTAINED FROM SIMPLE AND PARTIAL CORRELATION (MONTHLY OBSERVATIONS)

	Total	Average	Marginal
		Simple Regressions	
Combined cost	$cX_1 = 2935.59 + 1.998X_2$	$cX_1/X_2 = 1.998 + 2935.59/X_2$	$dcX_1/dX_2 = 1.998$
Productive labor cost	$pX_1 = -1695.16 + 1.780X_2$	$pX_1/X_2 = 1.780 - 1695.16/X_2$	$dpX_1/dX_2 = 1.780$
Nonproductive labor cost	$nX_1 = 992.23 + 0.097X_2$	$nX_1/X_2 = 0.097 + 992.23/X_2$	$dnX_1/dX_2 = 0.097$
Overhead cost	$oX_1 = 3638.30 + 0.121X_2$	$oX_1/X_2 = 0.121 + 3638.30/X_2$	$doX_1/dX_2 = 0.121$
		Partial Regressions	
Combined cost	$cX_1 = 762.54 + 2.068X_2$	$cX_1/X_2 = 2.068 + 762.54/X_2$	$dcX_1/dX_2 = 2.068$
Productive labor cost	$pX_1 = -2993.03 + 1.821X_2$	$pX_1/X_2 = 1.821 - 2993.03/X_2$	$dpX_1/dX_2 = 1.821$
Nonproductive labor cost	$nX_1 = 334.71 + 0.118X_2$	$nX_1/X_2 = 0.118 + 334.71/X_2$	$dnX_1/dX_1 = 0.118$
Overhead cost	$oX_1 = 3363.47 + 0.130X_2$	$oX_1/X_2 = 0.130 + 3363.47/X_2$	$doX_1/dX_2 = 0.130$

* The symbols have the following meaning:

cX_1 = combined cost in dollars oX_1 = overhead cost in dollars

pX_1 = productive labor cost in dollars X_2 = output in dozens of pairs

nX_1 = nonproductive labor cost in dollars

[3] The mathematics of this transformation may be illustrated by the following equations for monthly combined cost, where $_cX_1$ is total combined cost, and X_2 is output. The partial regression equation for combined cost in the form of monthly totals was found to be

$$_cX_1 = 762.54 + 2.068\, X_2$$

By dividing this equation through by X_2 the following equation for the combined cost per dozen was obtained:

$$\frac{_cX_1}{X_2} = 2.068 + \frac{762.54}{X_2}$$

By differentiating the total cost function with respect to X_2, the output, the resulting first derivative gives the marginal cost function as

$$\frac{d_cX_1}{dX_2} = 2.068.$$

[4] It should be remembered that these costs do not include raw material costs.

The graphic counterpart of the results obtained from the partial regression equation for combined cost expressed in Table 3 are shown in Chart 8. The upper section shows the partial regression of total monthly cost on output, which is the same as that shown in the upper section of Chart 4. The marginal cost function, which is pictured in the lower section, is the first derivative of this total cost function. Since the total cost function is linear, its slope obviously remains unchanged; hence marginal cost is constant. From these results it is seen that the operating cost of

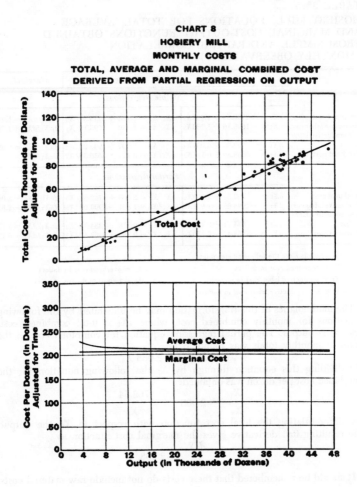

CHART 8
HOSIERY MILL
MONTHLY COSTS
TOTAL, AVERAGE AND MARGINAL COMBINED COST
DERIVED FROM PARTIAL REGRESSION ON OUTPUT

producing an additional dozen pairs of hosiery (not including the cost of silk) is approximately $2 over the range of output observed.[5]

The average cost function, which lies above the marginal cost line in the lower section, was obtained by dividing the total cost function by output (X_2). This curve shows how cost per dozen varies with the number of dozens produced. Since the fixed cost is relatively small compared to the variable cost, the average cost function is only slightly curved and lies very close to the marginal cost function.

[5] To be more precise, the marginal cost derived from the simple regression function is $1.998, while the esimate of marginal cost obtained from the partial regression function is $2.068.

Economies of Scale: Some Statistical Evidence

FREDERICK T. MOORE

Frederick Moore is a senior economist with the RAND
Corporation. This article appeared in the *Quarterly Journal
of Economics* in 1959.

I

Statistical evidence bearing on the existence of economies of scale in industry is, for the most part, sketchy and incomplete, although the logic of the economic and technical origins of such economies has been extensively developed. Reasons for this lack of statistical evidence are not hard to find; detailed cost studies of different sizes of plants are a sine qua non for analysis of the problem, yet such studies are difficult to obtain. Of necessity engineering information on technical possibilities for substitution among inputs must be combined with the mechanism of choice provided by economic calculations of cost. As Chenery has pointed out, the number of combinations of inputs that may be considered feasible by the engineer is much greater than the number observed in operation and studied by the economist; yet changes in relative prices alone will change the range of economically feasible combinations.

In lieu of deriving production functions from technical data (which is what is actually required), engineers—and in particular chemical engineers—have experimented with various "rules of thumb" for estimating the capital cost of plants of different sizes or for estimating process equipment costs. One such rule of thumb that has found some acceptance is the "0.6 factor" rule. The uses claimed and achieved for this rule will be summarized in a moment. Although the engineers do not seem to think of it as shedding light on economies of scale of plant, the rule can be so interpreted and will be discussed from that point of view.

Studies of capital coefficients (i.e., the ratio of capital expenditures to increases in capacity) by federal government agencies, universities, and others as part of an interindustry research program provide the statistical material for another evaluation of economies of scale. The methodology and results of these studies can be compared with those above.

II

The envelope cost curve usually serves as the vehicle for a discussion of economies of scale; the succession of plant short-run cost curves may trace out a smooth envelope curve or it may be scalloped in various ways. A discussion along this line overlooks the ways in which plant expansions actually take place, however. Expansions of capacity may occur through: the building of completely new plants at new locations: separate new productive facilities (multiple units) which utilize existing overhead facilities, such as office buildings, laboratories, etc.; the addition of new productive facilities which are intermingled with the old (the case of "scrambled" facilities); conversions of plants or processes from one product to another; or the elimination of "bottleneck" areas in a plant (the case of "unbalanced" expansion).

It is conceivable that the elimination of bottleneck areas in a plant will increase the capacity by a large amount (e.g., 50 percent); if that be the case, it is necessarily implied that in other areas of the plant there is excess capacity which can be utilized once the bottleneck is broken. This in turn implies that the productive units in the plant are not divisible, since, if they were, the plant could have been producing the old output with a smaller scale and lower costs. Thus it is usual to attribute economies of scale primarily—if not solely—to the lack of divisibility of productive units. Economies are realized by moving in the direction of larger common denominators of equipment, i.e., where fewer units are operated at less than capacity.

Size and equipment and indivisibilities therein are significant variables for a study of scale, but they do not necessarily go hand in hand. In a copper smelter, capacity may be increased by lengthening or widening the reverberatory furnace by small increments (thus increasing its cubic content). This ability to increase the size of a capital input by small amounts exists for a fairly wide selection of industrial equipment; in fact the usefulness of the ".6 rule" is really predicated on this occurrence. It has been noted by engineers that the cost of an item is frequently related to its surface area, while the capacity of the item increases in accordance with its volume. For that reason alone economies in scale may be achieved.

There is another matter that bears on this topic. Chamberlin has argued that it is not only divisibility but the aggregate amounts of inputs used that explain the existence of economies of scale. As size increases, the inputs change qualitatively as well as quantitatively. Different types of inputs are employed at various scales. Changes in quality mean changes in efficiency. The *form* of the input changes as well as the amount. It will not do to call this a question of classification, and to say that the inputs are really distinct. The functions performed by the inputs are the same; the quality changes do not alter the case.

In general it has been our experience in working with files of information on individual plant expansions in a number of industries that the complementary character of capital goods in a large expansion is quite marked. A large increase in capacity usually involves the plant in expenditures on all productive equipment, not just on selected items. This does not mean that fixed proportions are the rule; flexibility in the use of particular pieces of equipment is common. However, the isoquants probably tend to be more angular and less flat, as they would be in the case of easy substitution between inputs. (See the case of pipelines in Section IV for the opposite case.) Among other reasons, economies of scale arise because the proportions among inputs change as scale of plant changes, although the proportions are variable within certain limits. In other words, the "scale line" may have "kinks" in it as the size of the plant expands. The "kinks" indicate the points at which quality and quantity changes in inputs alter the proportions in which they tend to be used.

III

The ".6 rule" derived by the engineers is a rough method of measuring increases in capital cost as capacity is expanded. Briefly stated, the rule says that the increase in cost is given by the increase in capacity raised to the .6 power. Symbolically,

$$C_2 = C_1 \left(\frac{X_2}{X_1}\right)^{.6}$$

Here C_1 and C_2 are the costs of two pieces of equipment and X_1 and X_2 are their respective capacities. The rule has been adduced from the fact that for such items of equipment as tanks, gas holders, columns, compressors, the cost is determined by the amount of materials used in enclosing a given volume, i.e., cost is a function of surface area, while capacity is directly related to the volume of the container. Consider a spherical container. The area varies as the volume to the two-thirds power, or in other language, cost varies as capacity to the two-thirds power. If the container is cylindrical, then, by the same analogy, cost varies as capacity to the .5 power, if the volume is increased by changes in diameter, and if the ratio of height to diameter is kept constant, cost varies as capacity to the two-thirds power. From a consideration of these factors the .6 rule has been developed.

Now consider an alternative and generalized form of the .6 rule

$$E = aC^b$$

where E is capital expenditures, C is capacity, and a and b are parameters. As long as $b < 1$, there are economies in capital costs. These economies should not be interpreted as being identical with economies of scale since variable costs must also be considered in the latter case; however, there are some indications that labor, power, and utilities costs also decrease with increased scale while the costs of materials embodied in the final product remain constant. These indications are tentative and not demonstrated by statistical evidence in the cases that follow, so that the ensuing discussion on the evidences of economies of scale must be qualified.

Originally the .6 rule was applied to individual pieces of equipment or processes. A reasonable argument can be made for its validity in those cases; however, the regression line for the formula above cannot be indefinitely extrapolated. There are several reasons for this. In the first place an extrapolation of the line may lead to sizes of equipment that are larger than the standard sizes available or in which stresses beyond the limits of the material are involved. Nelson points out that, in building fractionating towers, an economical limit is reached at about 20-foot diameters since beyond that point very heavy beams are necessary in order to keep the trays level. Second, in some industries expansion takes place by a duplication of existing units rather than by an increase in their size, e.g., in aluminum reduction where several pot lines are constructed rather than enlarge individual pots. If the rule is to be applied at all, it is safest to limit its use to the range of capacities found in the observations.

The .6 rule when applied to complete plants runs into difficulties not encountered on individual equipment. Some expenditures are relatively fixed for large ranges of capacity, for example, the utilities system in the plant, the "overhead" facilities, plant transportation, instruments, etc. Complicated industrial machinery does not necessarily exhibit the same relationships between area (cost) and volume (capacity) as do simple structures like tanks and columns. Furthermore, for both items of equipment and complete plants, the gradations between sizes are not necessarily small. Indivisibilities in size are a real factor in some cases; an illustration from the crude pipeline industry will be discussed in Section IV.

In spite of these obvious limitations, estimates of the value of b in the formula

$$\log E = \log a + b \log C$$

have been made for a number of industries or products. These estimates are apt to be best for industries: (1) that are continuous-process rather than batch-operation; (2) that are capital-intensive; and (3) in which a homogeneous, standardized product is produced, so that problems of product-mix do not intrude to muddy the definition of capacity. The industries that best meet these criteria are the chemical industries (including petroleum), cement, and the milling, smelting, refining, and rolling and drawing of metals. These are the industries for which statistical estimates of b have been made, and for which some explanation of economies of scale has been supplied.

IV

Chilton has estimated values for b for 36 products in the chemical and metal industries. In three cases the value was greater than 1 but in only one of the cases was it so much larger as to be suspect. In the other 33 cases the values ranged from .48 to .91. The average value of b was .68 and the median .66, so that Chilton concluded that the .6 rule was reasonable even when extended to complete plants rather than individual pieces of equipment. Some of the values of b which Chilton obtained are shown in the following table. The petroleum industry is well represented in the sample; several processes and one example of complete refineries are shown.

Product	Value of b
Magnesium, ferrosilicon process	.62
Aluminum ingot	.90
TNT	1.01
Synthetic ammonia	.81
Styrene	.53
Aviation gasoline	.88
Complete refinery, including catalytic cracking	.75
Catalytic cracking, topping, feed preparation, gas recovery, polymerization	.88
Topping and thermal cracking	.60
Catalytic cracking	.81
Natural gasoline	.51
Thermal cracking	.62
Low purity oxygen	.47–.59

From the point of view of statistical appraisal of these results, it is unfortunate that the error in the regression equation and the standard error of b are not shown, although from a visual inspection of a few of the products it would appear that the correlations are very high. Nevertheless, it would be valuable to be able to apply a t-test to the b's to determine, for example, whether they differ significantly from 1. If they do not, the evidence on the existence of economies of scale in those industries would be shaky. It is reasonable to suppose that the values of b above .85 (approximately) are perhaps the ones most open to question.

The Harvard Economic Research Project directed by Professor Leontief has made estimates of these "scale factors" for a different selection of chemical products. Their results agree in general with those above, although the range of values found is greater (.2 to an aberrational value of 4.2), and the weighted average for 15 products is also higher than that found by Chilton. A selection of these values is as follows:

Product	Value of b
Aluminum sulfate from bauxite	4.2
Calcium carbide	.8
Carbon black, furnace process	.6
Carbon black, thermal decomposition	.2
Soda ash, Solvay process	.7
Styrene, from benzene and ethylene	.9
Sulfuric acid, contact process	.8
Synthetic rubber, Buna S	1.1

The average for 15 products (weighted by the United States Census values of shipments in 1947) was .8. The scale factors above were computed from very small samples. Of the 15 products studied, 8 were based on two observations; 2 were based on three observations; 2 on four observations; 1 on five and 2 on six. On the other hand, most of the observations were derived at least in part from engineering data, or were checked for type of process and completeness of design and equipment by engineers conversant with the industry. Nevertheless, the results must be viewed with skepticism. Furthermore, even in the cases in which there were the most observations (e.g., carbon black with six plants), the range of variation of equipment costs was considerable; the correlations do not appear to be very high. It is obvious that there are other factors such as location of the plant, product grade, etc., that affect capital expenditures; the data have not been adjusted to account for these factors so that the test of scale is not without ambiguity.

Under contract to the Bureau of Mines, the Petroleum Research Project, Rice Institute, has made a study of capital coefficients for crude oil and natural gas pipelines; one part of this study involved the derivation of a production function for pipelines and an investigation of economies of scale.

The two basic inputs of importance in the construction of a pipeline are the line pipe and the pumping stations, or, more accurately, the amount of hydraulic horsepower. The two inputs may be combined in a variety of ways to achieve any given capacity (which is defined as barrels per day of "throughput"). Any given throughput can be carried by substituting additional horsepower for a certain number of inches of (inside) diameter of pipe. Obviously, a pipe of smaller diameter involves less line pipe costs but also requires more expenditures on horsepower. For example, a throughput of 125,000 barrels per day (60 SUS oil over 1000 miles) can be obtained by any of the following combinations of pipe and horsepower:

(Outside) Diameter of pipe	Horsepower (approximate)
30	2,000
26	4,000
22	8,500
18	22,500
16	37,500

Other combinations of pipe diameter and hydraulic horsepower can be derived for throughputs greater or less than 125,000 barrels per day.

The isoquants relating diameter of pipe to hydraulic horsepower are of the usual form, convex to the origin, but they are relatively "flat," indicating a fairly easy substitution of these inputs for each other for any given throughput being considered.

Although the isoquants, in generalized form, appear as continuous curves which indicate that substitution possibilities may be considered in incremental amounts; in fact, there are discontinuities because pipe comes in standard sizes only. The most commonly used sizes for crude oil trunk lines have (outside) diameters of 8, 10, 12, 14, 16, 18, 20, 24, 26, and 30 inches. Inside diameters have a greater range of variation since wall thickness is also variable, but the number of sizes is not infinite; consequently, there are discontinuities in the production function.

The study of pipelines indicates clearly that economies of scale exist in the industry. Marginal physical product increases up to about 200,000 barrels per day, and for larger throughputs the marginal returns appear to be approximately constant. However, because of the discontinuities in the production function the line indicating increasing returns to scale may not cut the isoquants at points representing real alternatives in terms of line pipe size and horsepower. Furthermore, as the size of pumps increases, the cost per horsepower definitely decreases so that, although the marginal physical product tends to be constant above 200,000 barrels per day, the capital costs per unit may continue to fall if larger pumps are used. Although there are other costs to be considered, many of them are invariant with respect to throughput and are associated only with the length of the line so that they need not be considered for this problem.

V

Some selected industries in the minerals area have been studied using data obtained from records of plants built during World War II and during the mobilization period beginning in 1950. The records of the Defense Plant Corporation ("Plancors") and of applications of firms for rapid tax amortization ("TA's") contain information on specific expenditures for capital equipment and the increase in capacity which was expected. In order to obtain reasonably homogeneous data, observations selected for study were limited to completely new plants and large "balanced additions." Unbalanced expansions (the elimination of bottlenecks) were eliminated from consideration. This increase in sample homogeneity was thus accomplished at the expense of sample size; small samples were the rule rather than the exception. However, in partial compensation, each of the plants was studied intensively; the expenditures were classified by type and compared as between plants and processes within plants; in short, every precaution was taken qualitatively to increase the homogeneity of the data. In final form two statistics were presented for each plant: (1) the total capital expenditure (secured as the sum of individual expenditures on equipment and facilities); and (2) the capacity increase secured. These were then correlated using a linear function of the logarithms (i.e., in the form indicated above in this paper). The results in general corroborated those discussed above. In almost all cases the scale factor was less than 1. The industries covered are as follows.

A. Alumina:

Complete and detailed information was available on only two plants, both using the combination Bayer process for production of alumina; on both of the plants (Baton Rouge, Louisiana, and Hurricane Creek, Arkansas) the engineering designs and flow sheets as well as the engineering rated capacities were available. Scale factors for the complete plants and for particular process equipment in the plant were computed.

Plant or Equipment	Values of b
Total plant and equipment	.95
Total equipment	.93
Boiler shop products	.85
Construction and mining machinery	.24
Industrial furnaces and ovens	.98
Pipe and fittings	1.13

The value of b for the total plant corresponds closely to that secured by Harvard. The range of values secured for the process equipment is particularly interesting. The chief machinery complex in the plant exhibits very marked economies of scale, while the value for pipe and fittings indicates diseconomies of scale. It appears that the larger size plant (which has a yearly capacity of 778,000 tons compared with 500,000 tons for the other) can use machinery more efficiently but the connections among the units (piping, etc.) must become substantially more expensive in order, for example, to utilize fully a group of evaporators, mills, or filter presses. An analysis of the engineering flow diagram of the plant tends to confirm this deduction.

It also appears that short-run costs fall as output is expanded. Operating costs, including raw materials, operating labor, allocable share of overhead, and interest on working capital for the Baton Rouge plant have been estimated for three different levels of output.

Output	Operating Cost ($ per ton)
1000 tons/day	$27.28
500 tons/day	29.63
300 tons/day	32.43

B. Aluminum Reduction:

The sample consisted of eight plants comprising a little less than half of the total in existence. Some of the results of the calculations are summarized in the following table.

Item	b	Sy	r	σ_b
Total plant and equipment	.93	.038	.98	.06
Total equipment	.95	.021	.99	.03

A t test applied to the values of b, testing it against the hypothesis $b = 1$, gave values of 1.17 for total plant and equipment and 1.67 for equipment. Using a 5 percent critical probability level, neither of the values of b can be regarded as significantly different from 1, so that there is reason for questioning whether these values are really indicative of economies of scale in the industry.

This industry expands by introducing multiple pot lines rather than by an expansion in the size of individual process equipment, so that it is possible that the results would be improved if samples stratified according to number of pot lines were used. This suggests, of course, that there is a "lowest common denominator" for total equipment in the plant, and that the equipment is simply duplicated in any expansion, so that economies of scale cease once the lowest common denominator has been reached.

In this industry there are two basic processes of production that are basically smaller but which have different capital expenditures in certain process areas. In a prebaked carbon plant the carbon anodes are manufactured separately and then used in the pots; in Soderberg plants the carbon anodes are continuously replenished in the pot, so that expenditures on pot lines are larger. A Soderberg plant substitutes larger initial costs on equiment for lower operating costs; therefore a consideration of scale necessarily involves an attention to short-run operating costs in deciding on the type of plant to be built.

C. Aluminum Rolling and Drawing:

The sample in this industry consisted of four plants making rolled products and four making extrusions. The two types of operations were kept separate in the analysis. The results are summarized below:

Process	b	r	σ_b
Aluminum rolling			
Total plant	.88	.95	.16
Equipment	.81	.93	.18
Aluminum extrusions			
Total plant	1.00	.99	—
Equipment	.92	.97	.13

The t test applied to these results also fails to reveal values of b significantly different from 1; however, it is true here, as in aluminum reduction, that there are limits to the size of rolls or dies and that multiple units are the usual way in which capacity is expanded.

D. Cement:

The sample consisted of seven plants with a range in yearly capacity from 450,000 tons to 1,400,000 tons. For total plant the value for b was .77 and for equipment 1.06; the former value was not significantly different from 1 according to the t test.

The major variable in the construction of a cement plant is the size (length and diameter) of the kiln. Fuel economy in firing the kilns is a prime objective, since fuel constitutes a large part of operating costs. Kilns and allied furnace equipment may be almost infinitely varied in size; however, since the primary purpose of the kiln is the holding of a cubic charge, it was interesting to see if the .6 rule applied to kilns and to allied machinery in the cement plant.

| Construction and mining machinery | .60 |
| Furnaces and ovens (including kilns) | .73 |

These values accord well within the logic of the .6 rule.

E. Tonnage Oxygen:

The sample consisted of five plants ranging in capacity from 50 to 500 tons per day, and producing 95 percent oxygen. The value for b was .63. There are significant changes in capital inputs and costs in one process area (air compression) as scale increases. The major cost item in this area is compressors. For plants of up to 100 tons per day it is most economical to use reciprocating compressors, while between 100–200 tons, there is a choice of reciprocating or centrifugal compressors, and above 300 tons axial flow compressors are more economical. Not only the size, but, more particularly, the character of the capital input changes as the scale increases, and, since the horsepower-hours required per ton decrease as scale increases, there are distinct economies of scale in this process area of the plant. A value of b = .54 was computed for compressor types used in various sizes of plants.

VI

All of the above is but a smattering of evidence on the existence of economies of scale or the lack thereof. From a purely statistical point of view it is discouraging to find no scale factors that test out significantly against the hypothesis of constant returns; yet the samples are small, and above all it is not clear that a lack of homogeneity in the data does not vitiate the results. These are complex plants usually with a number of process areas. Some areas may be deliberately built with capacities larger than necessary in order to make easier any future expansion. If such is the case, the results are biased.

Although the formula may be applied to complete plants with useful result, it is clear that its application to particular

pieces of equipment or process areas is apt to provide better results. The statistical evidence is amply buttressed by engineering information on this point. By adhering strictly to processes
rather than complete plants, modifications in the formula can be
made to account for individual capacity-cost relationships. For
example, although a linear function of the logarithms seems to fit
most of the data well, there is some process equipment for which
a curvilinear function is required. For equipment such as cyclone
separators, centrifugals, and towers, a function that is concave
upward seems to fit the data better. In most cases these curves
indicate the existence of economies of scale up to a certain capacity (i.e., slope less than 1) and diseconomies beyond that
point (i.e., slope greater than 1); hence an average cost curve
for these items would turn up eventually but in general would
tend to be flat-bottomed over a considerable range in capacity.

Let us outline a general simple procedure for analyzing the
behavior of economies of scale using this process analysis. Suppose that plants in industry X can be divided into four main process areas and one "cooperating" area (e.g., the plant utilities
system, piping, or transportation); further let us assume that application of the formula to each area has produced the following
values for b:

Process area A	.25
Process area B	.60
Process area C	.80; 1.20
Process area D	1.00
Cooperating area E	1.10

From the table it is evident that there are economies of scale
in areas A, B, and C, although in the last the economies exist only
up to a certain point and then are replaced by diseconomies
(e.g., the fractionating tower mentioned previously). Area E
contains no possibilities for economies and area D provides constant returns to scale.

It would now be possible to investigate the behavior of economies of scale for different sizes of plant. Eventually the cost
curve may turn up. It depends on the importance (from the
standpoint of the percent of total expenditure) of areas C and
E. If 75 percent of total capital expenditures normally occur in

area C, or if the percent of expenditures in that area increases for larger sizes of plants, diseconomies of scale may occur fairly rapidly. If, on the other hand, area A is the most important in the plant, then economies of scale may continue over the whole observable range.

In order to assess the problem we should also know whether the scale of effort in each area can be expanded in small increments or whether the capacities of equipment increase by discrete amounts. In the latter case, economies of scale are limited to specific congeries of equipment. The qualitative characteristics of the equipment must also be investigated, since proportions may be affected thereby.

This would appear to be a relatively simple method of analyzing economies of scale in industry and one that is capable of use without an elaborate study of production functions. The engineers have compiled a good bit of information that can be used immediately, and catalogues of equipment can provide more. This information is not in the form that can be used directly; usually it specifies the cost of an item that can perform a certain job such as grinding a certain number of tons a day, or conveying a certain charge per hour, and so forth. But these data can be utilized with only small changes; three steps are normally involved:

1. The engineering data in technical journals and catalogues give cost relative to some engineering or physical magnitude (e.g., diameter of tank, square feet of heating surface, peripheral area, etc.).

2. The physical or engineering magnitude can be related to capacity by an appropriate formula (e.g., the capacity of a tank can be related to the diameter). Chenery has suggested some ways this can be done for whole processes, but what is suggested here is on a much simpler level; it may involve nothing more than an application of simple formula of area and volume, for example. Of course, in the process some of the elements may be omitted, but rough justice can usually be done to the relationship.

3. From (1) and (2) it is then possible to express the relationship between cost and capacity and to analyze the behavior of economies of scale.

It would be interesting to apply this procedure, process by process, to plants in several industries, to go through, in short, a simplified version of design of a plant including an analysis of the changes to be made in equipment as size varies. It would not be necessary to consider the whole range of substitutions among capital inputs that are possible; sufficient indications of economies of scale could be obtained from perhaps three or four typical sizes, so that the amount of analysis necessary would be smaller than for a complete production function analysis. It is hoped that others may find in this method much to commend as a simple procedure for evaluating the evidences of economies of scale.

SIMPLE REGRESSION: INCOME, EMPLOYMENT, AND CONSUMPTION

The purpose of the three articles in Part Five is to illustrate how simple regression can analyze relationships concerning income, employment, and consumption. All three of these articles are well known and important. The first, by Lawrence Klein and Richard Kosobud, uses simple regression to see whether some celebrated ratios of economics—the savings-income ratio, the capital-output ratio, labor's share of income, the income velocity of circulation, and the capital-labor ratio—have been subject to long-term upward or downward trends. The article provides some important basic data, in addition to illustrating the use of simple regression.

In the next paper, A. W. Phillips uses simple regression to describe the relationship between unemployment and the rate of change of money wage rates in the United Kingdom. He fits separate regressions for the period from 1861 to 1913, the period from 1913 to 1948, and the period from 1948 to 1967. With certain qualifications, he concludes that the evidence "seems in general to support the hypothesis . . . that the rate of change of money wage rates can be explained by the level of unemployment and the rate of change of unemployment . . ." The curves that are estimated have come to be known as Phillips curves, and they play an important, and controversial, role in modern economics.

The final paper, by Franco Modigliani, is concerned with the measurement of the consumption function. Modigliani begins by discussing several consumption functions that have been suggested by other economists.[1] Then he uses simple regression to estimate the relationship between the savings ratio and the difference (in percentage terms) between current income and the peak income in the past. Finally, he estimates a consumption function based on current income and peak income.

[1] Some parts of this article involve the use of multiple regression, but they should be understandable to readers who have studied only simple regression.

Great Ratios
Of Economics

LAWRENCE R. KLEIN and RICHARD F. KOSOBUD

Lawrence Klein is Benjamin Franklin Professor of
Economics at the Wharton School of the University of
Pennsylvania. Richard Kosobud teaches at Wayne State
University. This paper appeared in the *Quarterly Journal
of Economics* in 1961.

Economists frequently base their reasoning on key ratios
between variables. If these ratios are in the nature of fundamental parameters, simplifications of theory may result. If they are
simply ratios of variables, it is questionable whether any theoretical advances can be made through the transformation from statements about a quotient to statements about numerator and
denominator separately. Accountants often construct such key
ratios from quick assets and liabilities, or inventories and sales, or
earnings and fixed charges, and so forth. By reducing measurements for firms in diverse size groups to a common order of
magnitude, these ratios may be of use, as they are, in international
comparisons or historical growth comparisons. For theory construction, however, our standards must be high, and stability or
plainly systematic variation in ratios must be found in order to
enhance their usefulness.

Some celebrated ratios of economics are:

1. the savings-income ratio (S/Y),
2. the capital-output ratio (K/Y),
3. labor's share of income (wN/pY),
4. income velocity of circulation (pY/M),
5. the capital-labor ratio (K/N).

STATISTICAL ESTIMATION OF THE MODEL

With the abundance of long-run statistical series now available for the American economy, it may appear, *in prospect,* to be a fairly easy matter to collect the necessary series measuring each of the variables of the model in a mutually consistent fashion over a period as long as the first half of this century. It turns out, in fact, to be a substantial job of data collection and processing to prepare a consistent set of series for all the variables over this period.

The studies of Kuznets and Kendrick at the National Bureau of Economic Research are extremely helpful in providing series on national product, its components, and employment in a form that is readily adaptable to our uses.[1] The estimates of national product and capital formation by Kuznets are, in a sense, tailored to our needs by virtue of the facts that he provides series in both current and constant prices, that he revalues depreciation charges to replacement costs, and that he treats government capital formation like private capital formation. His estimates differ in one principal respect, however, from the official national accounting practices of the Department of Commerce. He does not classify total government expenditures on *current* goods and services as final purchases. He regards a large part of them (in recent years) as intermediate expenditures and excludes them from the total national product. He roughly allocates a small amount of them to personal consumption. In the postwar period of rapid expansion in the government sector, Kuznets' estimates are considerably lower than the official series. He gives a different statistical picture of the long-term growth of the American economy, and the estimates of our model reflect this fact.

Kuznets computes national product according to alternative variants. The one that we have selected shows no distinction between national income and net national product. It also allocates government expenditures, if they are not eliminated, to either consumption or investment spending. They are thus well suited to the global nature of our model in which our concepts

[1] S. Kuznets, *Capital in the American Economy: Its Formation and Financing* (New York: National Bureau of Economic Research, 1959, mimeographed). J. W. Kendrick, *Productivity Trends in the United States* (New York: National Bureau of Economic Research, 1960).

and accounting relations make no explicit allowance for institutional features of the government sector. Although his allocations or adjustments may be rough, Kuznets makes them definite on the product side of the accounts. In certain extreme periods, such as wartime, this leads to some anomalous results for our calculations unless some compensating adjustments are made on the income side.

THE SAVINGS–INCOME RATIO

Economists and statisticians have long been impressed with the findings of Kuznets that the percentage of income saved, by decades, has been fairly steady at about 10 percent for the period since the Civil War.[2] Goldsmith in his more recent massive study of savings has confirmed this result for the relationship between personal savings and personal income.[3] Kuznets' decade averages iron out cyclical fluctuations, but Goldsmith's annual estimates for this century exhibit strong cyclical influences about a steady trend. In the short run, the ratio is clearly not constant. In the long run, considering only the trend development, there is evidence of constancy in the personal savings-income ratio.

Our figures in the present paper differ from those of both studies cited above. We examine the constancy of the ratio between consumption and net national product for the period 1900–1953 as computed by Kuznets in his more recent study. Both series are expressed in 1929 prices. Since we deal with the consumption-income ratio directly for measurement purposes, the savings-income ratio is only indirectly measured as a residual. The implied concept of savings includes more than personal savings. It includes business and government savings as well. The income or product concept measures national income and not personal income.

Alternative probability structures underlying the estimates of the savings ratio, α, are plausible. They might be

[2] S. Kuznets, *Uses of National Income in Peace and War* (New York: National Bureau of Economic Research, 1942), p. 30.
[3] R. W. Goldsmith, *A Study of Savings in the United States*, Vols. I, III (Princeton, N. J.: Princeton University Press, 1955, 1956).

$$C/Y = (1 - \alpha) + u;$$
$$C/Y = (1 - \alpha)u;$$
$$C = (1 - \alpha)Y + u;$$

where C = consumption; Y = national product and u = random error.

In the first case, $(1 - \alpha)$ would be estimated as the arithmetic mean of C/Y, and in the second as the geometric mean. In the third case, some form of unbiased regression estimates would be needed. We have made both of the first two types of estimates. Our charts plot the ratios in arithmetic units. Our computed equations are presented in logarithmic units, suited to the second case.[4]

The series are given in the accompanying table and a chart of the consumption-income ratio together with the other great ratios is given in Fig. 1. The data show a consumption-income ratio just under 0.9 at the beginning of this century and a series of values above this level in recent years. A significant upward trend is suggested for this series. Our estimate is

$$\log \frac{C}{Y} = -0.03933 + 0.00054 \, t$$

or

$$\frac{C}{Y} = 0.9134 \, (1.00129)^t$$

Time, t, is measured chronologically, centered at January 1, 1927, in terms of six-months units. At the midpoint of the sample span, the ratio is 0.9134 and increases at a compound interest rate of about 1/8 of 1 percent semiannually. This trend is statistically significant. In the logarithmic formulation, the coefficient of t is more than five times as large as its sampling error.

We conclude that the savings-income ratio is not a constant, but is on a *declining* trend (opposite trend to that of the consumption-income ratio).[5]

[4] Both of the first two forms have been computed. The numerical results for the second are analyzed in the text.

[5] Compare the observations of G. D. A. MacDougall, "Does Productivity Rise Faster in the United States?" *The Review of Economics and Statistics*, 38 (May 1956), 173.

THE CAPITAL–OUTPUT RATIO

In the currently fashionable economics of growth, theorists do not rigidly assume constancy of the accelerator coeffi-

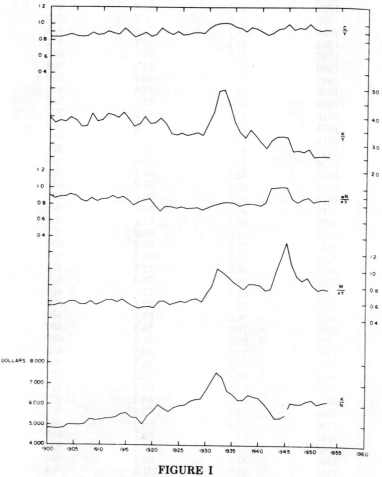

FIGURE I
THE GREAT RATIOS OF ECONOMICS

TABLE 1

CONSUMPTION AND NET NATIONAL PRODUCT, UNITED STATES
(BILLIONS OF 1929 DOLLARS)

	Consumption	Net National Product	Ratio C/Y
1900	27.8	33.0	.842
1901	30.5	36.3	.840
1902	30.9	36.8	.840
1903	32.9	38.8	.848
1904	33.3	38.1	.874
1905	34.9	40.7	.857
1906	38.3	45.4	.844
1907	39.7	46.7	.850
1908	38.1	42.6	.894
1909	41.4	47.8	.866
1910	42.1	48.2	.873
1911	43.2	47.9	.902
1912	42.8	48.5	.882
1913	44.7	51.3	.871
1914	47.1	49.8	.946
1915	48.2	53.7	.898
1916	49.4	58.8	.840
1917	50.8	59.0	.861
1918	49.6	55.4	.895
1919	52.2	61.1	.854
1920	54.2	62.2	.871
1921	57.0	59.6	.956
1922	59.2	63.9	.926
1923	64.3	73.5	.875
1924	69.0	75.6	.913
1925	67.1	77.3	.868
1926	72.5	82.8	.876
1927	74.2	83.6	.888
1928	76.3	84.9	.899
1929	80.3	90.3	.889
1930	75.9	80.5	.943
1931	73.2	73.5	.996
1932	66.4	60.3	1.101
1933	65.0	58.2	1.117
1934	68.6	64.4	1.065
1935	73.1	75.4	.969
1936	80.8	85.0	.951
1937	84.4	92.7	.910
1938	83.0	85.4	.972
1939	87.0	92.3	.943
1940	91.7	101.2	.906
1941	97.9	113.3	.864
1942	96.2	107.8	.892
1943	98.8	105.2	.939
1944	102.2	107.1	.954

TABLE 1 (CONT.)
CONSUMPTION AND NET NATIONAL PRODUCT, UNITED STATES
(BILLIONS OF 1929 DOLLARS)

	Consumption	Net National Product	Ratio C/Y
1945	109.1	108.8	1.003
1946	122.3	131.4	.931
1947	124.9	130.9	.954
1948	127.5	134.7	.947
1949	130.7	129.1	1.012
1950	138.7	147.8	.938
1951	139.8	152.1	.919
1952	143.9	154.3	.933
1953	149.4	159.9	.934

Source: Kuznets, *Capital in the American Economy.*

cient. In the short run, this ratio may be more constant if measured with capacity instead of actual output. In the long run, it is likely to fall in advanced industrial economies as a result of technical progress.

As in the case of the savings-income ratio, two of the principal investigators of the statistical capital-output ratio have been Kuznets and Goldsmith.[6] Kuznets' decade estimates rise from 2.83 to 3.19 between 1879 and 1944. There are great swings within this period, however. Goldsmith's annual estimates confirm this general pattern until World War II, after which his ratio shows a tendency to fall. Our estimate for this century, obtained by cumulating Kuznets' annual net investment figures from a starting figure for capital stock, shows a significant downward trend. (Table 2 and Fig. 1) Our equation is

$$\log \frac{K}{Y} = 0.54699 - 0.0015t$$

or

$$\frac{K}{Y} = 3.523 \ (1.0033)^{-t}$$

[6] S. Kuznets, "Long-Term Changes in the National Product of the United States of America since 1870"; R. W. Goldsmith, "The Growth of Reproducible Wealth of the United States of America from 1805 to 1905," *Income and Wealth*, Series II (Cambridge: Bowes and Bowes, 1952).

TABLE 2
CAPITAL STOCK AND NET NATIONAL PRODUCT, UNITED STATES
(BILLION OF 1929 DOLLARS)

	Capital Stock	Net National Product	Ratio K/Y
1900	131.57	33.0	3.987
1901	136.71	36.3	3.766
1902	142.27	36.8	3.866
1903	147.72	38.8	3.807
1904	152.20	38.1	3.995
1905	157.65	40.7	3.873
1906	164.50	45.4	3.623
1907	171.25	46.7	3.667
1908	175.31	42.6	4.115
1909	182.05	47.8	3.809
1910	187.86	48.2	3.898
1911	192.45	47.9	4.108
1912	198.04	48.5	4.083
1913	204.38	51.3	3.984
1914	207.28	49.8	4.162
1915	210.36	53.7	3.917
1916	215.70	58.8	3.668
1917	220.26	59.0	3.733
1918	223.94	55.4	4.042
1919	229.37	61.1	3.754
1920	235.13	62.2	3.780
1921	236.19	59.6	3.963
1922	240.15	63.9	3.758
1923	248.87	73.5	3.386
1924	254.47	75.6	3.336
1925	264.08	77.3	3.416
1926	273.94	82.8	3.308
1927	282.61	83.6	3.381
1928	290.20	84.9	3.418
1929	299.42	90.3	3.316
1930	303.30	80.5	3.768
1931	303.42	73.5	4.128
1932	297.12	60.3	4.927
1933	290.12	58.2	4.985
1934	285.43	64.4	4.432
1935	287.78	75.4	3.817
1936	292.11	85.0	3.437
1937	300.33	92.7	3.240
1938	301.38	85.4	3.529
1939	305.58	92.3	3.311
1940	313.32	101.2	3.096
1941	327.41	113.3	2.890
1942	338.98	107.8	3.145
1943	347.08	105.2	3.299
1944	353.46	107.1	3.300
1945	354.13	108.8	3.255

TABLE 2 (CONT.)
CAPITAL STOCK AND NET NATIONAL PRODUCT, UNITED STATES
(BILLION OF 1929 DOLLARS)

	Capital Stock	Net National Product	Ratio K/Y
1946	359.43	131.4	2.735
1947	359.30	130.9	2.745
1948	365.19	134.7	2.711
1949	363.21	129.1	2.813
1950	373.73	147.8	2.529
1951	385.95	152.1	2.537
1952	396.47	154.3	2.569
1953	408.01	159.9	2.552

Source: Kuznets, *Capital in the American Economy.*

This equation puts the rate of decline at about ⅓ of 1 percent semiannually. At the midpoint, $t = 0$, we have a ratio of 3.523, and the coefficient of t in the logarithmic equation is seven times its estimated sampling error.

We did not estimate the accelerator form of this equation directly. To do so would require a different set of assumptions about the probability structure of the model and would involve the use of negative numbers, thus precluding the estimation of logarithmic trends.

LABOR'S SHARE

The wage share of national income has received at least as much measurement attention as any of the great ratios. In this case much more experimentation has been made with alternative numerators and denominators. Should payments to labor include salaries generally, salaries of company executives, income of small proprietors, or employer contributions to retirement? The denominator may be gross product, net product, national income, or personal income. Dunlop has charted labor's share for a great variety of these alternative concepts.[7] His graphs show no statistical grounds for choosing among alternative formulations. Over the long run, it appears, from his data, that the

[7] J. T. Dunlop, *Wage Determination under Trade Unions* (New York: Kelley, (1950), Chap. 8.

ratios are stable with considerable cyclical fluctuation. His findings are different for wages and for salaries. They also differ among industry groups, but exhibit no trend for the economy as a whole during the interwar period.

Johnson has imputed a wage to self-employed persons.[8] From the first decade of this century to the post World War II period he finds a growth in labor's share of about 5 percentage points. This longer period gives different results from Dunlop's interwar period. Were Johnson not to impute wages to self-employed persons, he would find nearly twice as large an increase in labor's share.

Kravis, in a study of Johnson's and later figures, adduces reasons for an increasing trend in labor's share.[9] He argues that demand for both labor and capital increased with output growth in an expanding economy. Capital supply was more responsive to the increased demand, and the comparative inelasticity of labor's response raised the wage share in national income.

The shift in population from rural to urban areas, and the increasing importance of the government sector may account for a large part of the increase in the share paid to labor, since both phenomena represent a growth in sectors paying a larger share of output to labor. Social welfare legislation and growing strength of trade unions may also account for some of the increase.

If one confines calculations to what is purely wage income in official series, an increasing trend is likely to result. This will be true for the nonfarm private sector as well as for the whole economy; hence industry shifts will not fully explain the trend. Noticing, however, that the trend was reduced in Johnson's series if imputations of wage income to self-employed persons are made, and following a suggestion in Kuznets' work, we combine the whole of self-employed income with wage and salary income. The total of such income from active employment tends to be nearly a constant fraction of national income as far as trends are concerned. This wider scope of wage income is consistent with

[8] D. G. Johnson, "The Functional Distribution of Income in the United States, 1950–1952," *Review of Economics and Statistics,* 36 (May 1954) 175–82.
[9] I. B. Kravis, "Relative Income Shares in Fact and Theory," *American Economic Review,* 49 (December 1959), 917–49.

our scope for employment in the model, which is to include all persons engaged, whether they be production workers, executives, managers, or self-employed. This wider scope of the payments and employment series is consistent also with the global character of our model.

Weintraub, in a recent study of price level phenomena, makes strong use of the constancy of labor's share. His numerator is the total of private wages and salaries. His denominator is the private gross national product. He argues in favor of constancy of this ratio from 1929 to recent years.[10]

Our data consist of the present official series published by the Department of Commerce on employee compensation and income from self-employment, extended back from 1929 to 1900 by splicing to Johnson's corresponding series. This series is expressed in current prices. The denominator of our ratio is the current price value of net national product estimated by Kuznets. From Table 3, we can see the curious result that during World War II the earnings variable exceeded Kuznets' adjusted value of national product.[11]

A formal calculation of the trend in the ratio yields

$$\log \frac{wN}{pY} = -0.07369 + 0.000082t$$

or

$$\frac{wN}{pY} = 0.8439 \, (1.00019)^t.$$

[10] S. Weintraub, *A General Theory of the Price Level, Output, Income Distribution and Economic Growth* (Philadelphia: Chilton, 1959).

[11] From 1941 an adjustment had to be made to the income side of the national accounts to correspond with Kuznets' adjustments to the product side. He subtracted personal tax and nontax payments from consumption (except for 3.6 percent of consumption, which he treated as *final* government services to consumers). We subtracted this amount from the numerator of our ratio. There are obviously inaccuracies in this rough adjustment, but it brings us close to the Bowman-Easterlin treatment of the income side for Kuznets' concepts. They argue for factorial imputations of income after taxes. See R. Bowman and R. A. Easterlin, "The Income Side: Some Theoretical Aspects," in *A Critique of the U. S. Income and Product Accounts*, National Bureau of Economic Research (Princeton: Princeton University Press, 1958), pp. 180–86.

TABLE 3
EARNED INCOME AND NET NATIONAL PRODUCT,
UNITED STATES
(BILLIONS OF CURRENT DOLLARS)

	Earned Income	Net National Product	Ratio wN/pY
1900	14.9	16.4	.909
1901	15.8	18.0	.878
1902	16.8	18.8	.894
1903	17.9	20.0	.895
1904	18.3	19.9	.920
1905	19.5	21.7	.899
1906	21.0	24.8	.847
1907	22.1	26.5	.834
1908	21.1	24.1	.876
1909	23.6	28.1	.840
1910	24.9	29.0	.859
1911	24.7	28.6	.864
1912	27.2	30.2	.901
1913	27.9	32.1	.869
1914	28.3	31.6	.896
1915	30.1	35.2	.855
1916	34.3	43.5	.789
1917	44.5	53.9	.826
1918	49.5	58.2	.851
1919	56.5	65.2	.867
1920	59.6	75.7	.787
1921	44.4	61.8	.718
1922	48.7	63.0	.773
1923	57.0	74.1	.769
1924	56.6	75.2	.753
1925	60.8	78.6	.774
1926	63.2	84.6	.747
1927	63.0	83.1	.758
1928	64.6	85.0	.760
1929	65.9	90.3	.730
1930	58.3	76.9	.758
1931	48.4	61.7	.784
1932	36.4	44.8	.812
1933	35.1	42.6	.824
1934	41.3	50.3	.821
1935	47.7	58.2	.820
1936	53.4	68.2	.782
1937	60.6	75.1	.807
1938	56.1	68.8	.815
1939	59.7	73.8	.809
1940	65.1	81.8	.796
1941	82.2	99.0	.826
1942	109.2	106.5	.999
1943	137.8	113.2	1.092
1944	150.9	118.7	1.146

TABLE 3 (CONT.)
EARNED INCOME AND NET NATIONAL PRODUCT,
UNITED STATES
(BILLIONS OF CURRENT DOLLARS)

	Earned Income	Net National Product	Ratio wN/pY
1945	154.0	124.2	1.107
1946	154.3	160.5	.877
1947	164.3	179.0	.831
1948	181.2	192.8	.863
1949	176.4	185.8	.884
1950	191.7	215.0	.827
1951	222.3	238.1	.844
1952	237.2	245.6	.858
1953	249.5	258.8	.858

Source: Department of Commerce; Kuznets, *Capital in the American Economy,* and Johnson, "The Functional Distribution of Income in the United States, 1850–1952."

The coefficient of t in the logarithmic form is less than half its sampling error. We suggest that there is no trend in this ratio. Without the trend we have

$$\frac{wN}{pY} = 0.8439$$

VELOCITY OF CIRCULATION

It is an interesting property of the present system that the savings ratio, the accelerator coefficient, labor's share, and velocity can all be put together in a consistent framework, for separate investigators who have searched for economic insight in terms of any single one of these ratios have often been in heated dispute with one another. It is especially true that the velocity analysts have been set apart as students of monetary phenomena with entirely different views from those concerned with "real" phenomena. The velocity ratio, by itself, has received great attention in a variety of forms, depending on the choice of numerator and denominator. Cash balances in the numerator may cover only checking accounts and circulating currency, or may be expanded to include time deposits, savings and loan shares, savings bonds, and other "near" moneys. Balances may be segregated according as they are held by persons, business, or

financial institutions. The denominator could be national expenditure, national income, personal income, disposable income, or
some broad duplicative measure of transactions.

A recent authoritative summary of velocity statistics computed by many authors is given by Selden.[12] For his own calculations, Selden defines cash to include total deposits, currency outside banks, Treasury deposits with Federal Reserve banks, and
money held in the Treasury. His denominator is measured as
national income. He estimates velocity in a range between 0.75
and 1.76 for annual periods in this half-century. He finds both a
trend and a cycle in these estimates.

Our estimates are based on a money total that includes circulating currency, demand deposits, and time deposits of persons,
business and government. The denominator of our ratio is the
current dollar value of net national product, according to Kuznets' variant that does not differ from national income.

As Table 4 shows, there is a noticeable trend in this series.
Our estimate is

$$\log \frac{M}{pY} = -0.15500 + 0.0025t$$

or

$$\frac{M}{pY} = 0.6998 \ (1.0057)^t.$$

At the midpoint of our series, the estimate of the reciprocal of
velocity is 0.6998. This represents the well-known Cambridge k.
In the semilogarithmic form that was fitted to the data, the coefficient of t is more than ten times its sampling error.

THE CAPITAL–LABOR RATIO

The capital-output ratio and the ratio of output to employment (productivity) have long been studied in great detail.
Together they define the capital-labor ratio, but in this form
the statistics have not so frequently been investigated. Kuz-

12 R. T. Selden, "Monetary Velocity in the United States," *Studies in the
Quantity Theory of Money,* ed. M. Friedman (Chicago: University of Chicago Press, 1956), pp. 179–257.

TABLE 4

CASH BALANCES AND NET NATIONAL PRODUCT,
UNITED STATES
(BILLIONS OF CURRENT DOLLARS)

	Cash Balances	Net National Product	Ratio M/pY
1900	8.9	16.4	.543
1901	10.0	18.0	.556
1902	10.8	18.8	.574
1903	11.5	20.0	.575
1904	12.0	19.9	.603
1905	13.2	21.7	.608
1906	14.1	24.8	.569
1907	15.1	26.5	.570
1908	14.7	24.1	.610
1909	15.8	28.1	.562
1910	17.0	29.0	.586
1911	17.8	28.6	.622
1912	18.9	30.2	.626
1913	19.4	32.1	.604
1914	20.0	31.6	.633
1915	20.7	35.2	.588
1916	24.2	43.5	.556
1917	28.2	53.9	.523
1918	31.4	58.2	.540
1919	35.6	65.2	.546
1920	39.9	75.7	.527
1921	37.8	61.8	.612
1922	39.0	63.0	.619
1923	42.7	74.1	.576
1924	44.5	75.2	.592
1925	48.3	78.6	.615
1926	50.6	84.6	.598
1927	52.2	83.1	.628
1928	54.7	85.0	.644
1929	55.2	90.3	.611
1930	54.4	76.9	.707
1931	52.9	61.7	.857
1932	45.4	44.8	1.013
1933	41.7	42.6	.975
1934	46.0	50.3	.915
1935	49.9	58.2	.857
1936	55.1	68.2	.807
1937	57.3	75.1	.763
1938	56.6	68.8	.823
1939	60.9	73.8	.825
1940	67.0	81.8	.819
1941	74.2	99.0	.749
1942	82.0	106.5	.770
1943	110.2	113.2	.973
1944	136.2	118.7	1.147

TABLE 4 (CONT.)
CASH BALANCES AND NET NATIONAL PRODUCT,
UNITED STATES
(BILLIONS OF CURRENT DOLLARS)

	Cash Balances	Net National Product	Ratio M/pY
1945	162.8	124.2	1.311
1946	171.2	160.5	1.067
1947	165.5	179.0	.925
1948	167.9	192.8	.871
1949	167.9	185.8	.904
1950	173.8	215.0	.808
1951	181.0	238.1	.760
1952	191.0	245.6	.778
1953	197.6	258.8	.762

Source: Board of Governors of the Federal Reserve System; Kuznets, *Capital in the American Economy.*

nets has, however, analyzed this ratio in his long-run studies of the American economy.[13] He estimates that the capital-labor ratio has nearly tripled between 1879 and 1944. The most rapid growth occurred at the turn of the century.

Our series show a steady upward growth in this ratio from about $5000 per person engaged at about 1900 to about $6000 after World War II. This amounts to a compound interest rate of growth slightly under ¼ of 1 percent semiannually. Our trend formulas are

$$\log \frac{K}{N} = 3.76126 + 0.0010t$$

or

$$\frac{K}{N} = 5571 \ (1.0023)^t.$$

The linear trend coefficient is statistically significant. It is more than six times its estimated sampling error. The series are given in Table 5 and Fig. 1.

[13] S. Kuznets, "Long Term Changes in the National Product of the United States of America Since 1870."

TABLE 5

CAPITAL STOCK AND PERSONS ENGAGED, UNITED STATES
(BILLIONS OF 1929 DOLLARS AND MILLIONS OF PERSONS)

	Capital Stock	Persons Engaged	Ratio K/N
1900	131.57	27.3	4820
1901	136.71	28.4	4814
1902	142.27	29.6	4806
1903	147.72	30.5	4843
1904	152.20	30.4	5007
1905	157.65	31.8	4958
1906	164.50	33.1	4970
1907	171.25	33.8	5067
1908	175.31	33.1	5296
1909	182.05	34.8	5231
1910	187.86	35.7	5262
1911	192.45	36.3	5302
1912	198.04	37.3	5309
1913	204.38	37.9	5393
1914	207.28	37.5	5527
1915	210.36	37.7	5580
1916	215.70	40.1	5379
1917	220.26	41.5	5307
1918	223.94	44.0	5090
1919	229.37	42.3	5422
1920	235.13	41.5	5666
1921	236.19	39.4	5995
1922	240.15	41.4	5801
1923	248.87	43.9	5669
1924	254.47	43.3	5877
1925	264.08	44.5	5934
1926	273.94	45.8	5981
1927	282.61	45.9	6157
1928	290.20	46.4	6254
1929	299.42	47.6	6290
1930	303.30	45.5	6666
1931	303.42	42.6	7123
1932	297.12	39.3	7560
1933	290.12	39.6	7326
1934	285.43	42.7	6685
1935	287.78	44.2	6511
1936	292.11	47.1	6202
1937	300.33	48.2	6231
1938	301.38	46.4	6495
1939	305.58	47.8	6393
1940	313.32	49.6	6317
1941	327.41	54.1	6052
1942	338.98	59.1	5736
1943	347.08	64.9	5348
1944	353.46	66.0	5355
1945	354.13	64.4	5499
1946	359.43	58.9	6102

TABLE 5 (CONT.)
CAPITAL STOCK AND PERSONS ENGAGED, UNITED STATES
(BILLIONS OF 1929 DOLLARS AND MILLIONS OF PERSONS)

	Capital Stock	Persons Engaged	Ratio K/N
1947	359.30	59.3	6059
1948	365.19	60.2	6066
1949	363.21	58.7	6188
1950	373.73	60.0	6228
1951	385.95	63.8	6049
1952	396.47	64.9	6109
1953	408.01	66.0	6182

Source: Kuznets, *Capital in the American Economy,* and Kendrick, *Productivity Trends in the United States.*

The Relation Between Unemployment and the Rate of Change of Money Wage Rates in the United Kingdom, 1861-1957

A. W. PHILLIPS

A. W. Phillips is Professor of Economics at the Australian National University. This article appeared in *Economica* in 1958.

1. HYPOTHESIS

When the demand for a commodity or service is high relative to the supply of it, we expect the price to rise, the rate of rise being greater the greater the excess demand. Conversely, when the demand is low relative to the supply, we expect the price to fall, the rate of fall being greater the greater the deficiency of demand. It seems plausible that this principle should operate as one of the factors determining the rate of change of money wage rates, which are the price of labor services. When the demand for labor is high and there are very few unemployed, we should expect employers to bid wage rates up quite rapidly, each firm and each industry being continually tempted to offer a little above the prevailing rates to attract the most suitable labor from other firms and industries. On the other hand, it appears that workers are reluctant to offer their services at less than the prevailing rates when the demand for labor is low and unemployment is high so that wage rates fall only very slowly. The relation between unemployment and the rate of change of wage rates is therefore likely to be highly nonlinear.

It seems possible that a second factor influencing the rate of change of money wage rates might be the rate of change of the demand for labor, and so of unemployment. (Thus in a year of rising business activity, with the demand for labor increasing and the percentage of unemployment decreasing, employers will be bidding more vigorously for the services of labor than they would be in a year during which the average percentage of unemployment was the same but the demand for labor was not increasing.) Conversely in a year of falling business activity, with the demand for labor decreasing and the percentage of unemployment increasing, employers will be less inclined to grant wage increases, and workers will be in a weaker position to press for them, than they would be in a year during which the average percentage of unemployment was the same but the demand for labor was not decreasing.

A third factor that may affect the rate of change of money wage rates is the rate of change of retail prices, operating through cost of living adjustments in wage rates. It will be argued here, however, that cost of living adjustments will have little or no effect on the rate of change of money wage rates except at times when retail prices are forced up by a very rapid rise in import prices (or, on rare occasions in the United Kingdom, in the prices of home-produced agricultural products). For suppose that productivity is increasing steadily at the rate of, say, 2 percent per annum and that aggregate demand is increasing similarly so that unemployment is remaining constant at, say, 2 percent. Assume that with this level of unemployment and without any cost of living adjustments wage rates rise by, say, 3 percent per annum as the result of employers' competitive bidding for labor, and that import prices and the prices of other factor services are also rising by 3 percent per annum. Then retail prices will be rising on average at the rate of about 1 percent per annum (the rate of change of factor costs minus the rate of change of productivity). Under these conditions the introduction of cost of living adjustments in wage rates will have no effect, for employers will merely be giving under the name of cost of living adjustments part of the wage increases which they would in any case have given as a result of their competitive bidding for labor.

Assuming that the value of imports is one-fifth of national income, it is only at times when the annual rate of change of import prices exceeds the rate at which wage rates would rise as a result of competitive bidding by employers, by more than five times the rate of increase of productivity, that cost of living adjustments become an operative factor in increasing the rate of change of money wage rates. Thus in the example given a rate of increase of import prices of more than 13 percent per annum would more than offset the effects of rising productivity, so that retail prices would rise by more than 3 percent per annum. Cost of living adjustments would then lead to a greater increase in wage rates than would have occurred as a result of employers' demand for labor, and this would cause a further increase in retail prices, the rapid rise in import prices thus initiating a wage-price spiral which would continue until the rate of increase of import prices dropped significantly below the critical value of about 13 percent per annum.

The purpose of the present study is to see whether statistical evidence supports the hypothesis that the rate of change of money wage rates in the United Kingdom can be explained by the level of unemployment and the rate of change of unemployment, except in or immediately after those years in which there was a very rapid rise in import prices, and if so to form some quantitative estimate of the relation between unemployment and the rate of change of money wage rates. The periods 1861–1913, 1913–48 and 1948–57 will be considered separately.

2. 1861–1913

Schlote's index of the average price of imports [1] shows an increase of 12.5 percent in import prices in 1862 as compared with the previous year, an increase of 7.6 percent in 1900 and in 1910, and an increase of 7.0 percent in 1872. In no other year between 1861 and 1913 was there an increase in import prices of as much as 5 percent. If the hypothesis stated above is correct, the rise in import prices in 1862 may just have been sufficient to start up a mild wage-price spiral, but in the remainder of the period changes in import prices will have had little or no effect on the rate of change of wage rates.

A scatter diagram of the rate of change of wage rates and the

[1] W. Schlote, *British Overseas Trade from 1700 to the 1930's*, Table 26.

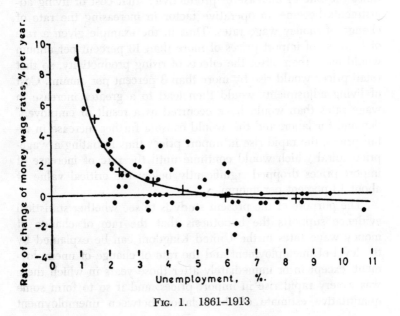

FIG. 1. 1861–1913

percentage unemployment for the years 1861–1913 is shown in Fig. 1. During this time there were 6½ fairly regular trade cycles with an average period of about eight years. Scatter diagrams for the years of each trade cycle are shown in Figs. 2 to 8. Each dot in the diagrams represents a year, the average rate of change of money wage rates during the year being given by the scale on the vertical axis and the average unemployment during the year by the scale on the horizontal axis. The rate of change of money wage rates was calculated from the index of hourly wage rates constructed by Phelps Brown and Sheila Hopkins,[2] by expressing the first central difference of the index for each year as a percentage of the index for the same year. Thus the rate of change for 1861 is taken to be half the difference between the index for 1862 and the index for 1860 expressed as a percentage of the

[2] E. H. Phelps Brown and Sheila Hopkins, "The Course of Wage Rates in Five Countries, 1860–1939," *Oxford Economic Papers,* June 1950.

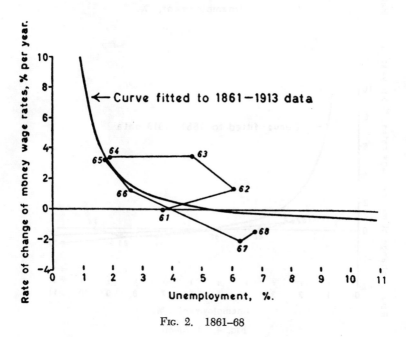

FIG. 2.   1861–68

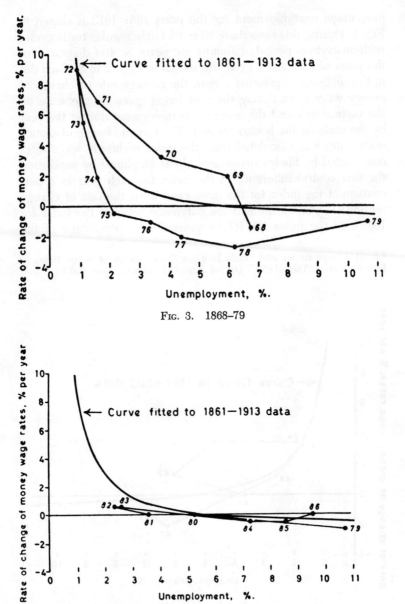

Fɪɢ. 3.   1868–79

Fɪɢ. 4.   1879–86

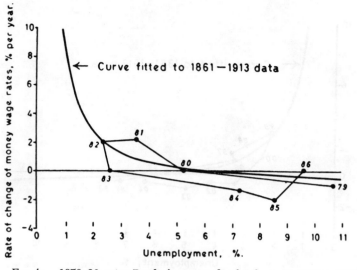

FIG. 4a. 1879–86, using Bowley's wage index for the years 1881 to 1886

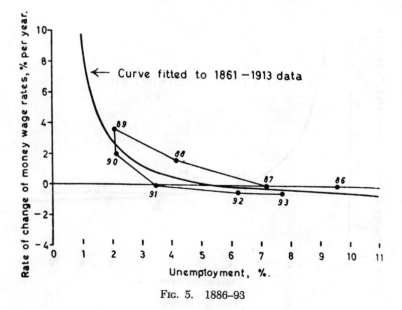

FIG. 5. 1886–93

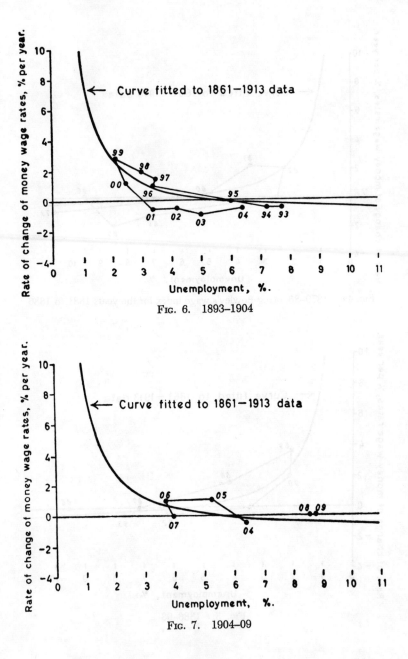

FIG. 6.   1893–1904

FIG. 7.   1904–09

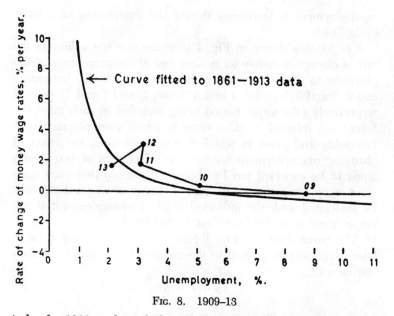

FIG. 8. 1909-13

index for 1861, and similarly for other years.[3] The percentage unemployment figures are those calculated by the Board of Trade and the Ministry of Labor from trade union returns. The corresponding percentage employment figures are quoted in Beveridge, *Full Employment in a Free Society*, Table 22.

It will be seen from Figs. 2 to 8 that there is a clear tendency for the rate of change of money wage rates to be high when unemployment is low and to be low or negative when unemployment is high. There is also a clear tendency for the rate of change of money wage rates at any given level of unemployment to be above the average for that level of unemployment when unemployment is decreasing during the upswing of a trade cycle, and to be below the average for that level of unemployment when

---

[3] The index is apparently intended to measure the average of wage rates during each year. The first central difference is therefore the best simple approximation to the average absolute rate of change of wage rates during a year, and the central difference expressed as a percentage of the index number is an appropriate measure of the average percentage rate of change of wage rates during the year.

unemployment is increasing during the downswing of a trade cycle.

The crosses shown in Fig. 1 give the average values of the rate of change of money wage rates and of the percentage unemployment in those years in which unemployment lay between 0 and 2, 2 and 3, 3 and 4, 4 and 5, 5 and 7, and 7 and 11 percent respectively (the upper bound being included in each interval). Since each interval includes years in which unemployment was increasing and years in which it was decreasing, the effect of changing unemployment on the rate of change of wage rates tends to be canceled out by this averaging, so that each cross gives an approximation to the rate of change of wages that would be associated with the indicated level of unemployment if unemployment were held constant at that level.

The curve shown in Fig. 1 (and repeated for comparison in later diagrams) was fitted to the crosses. The form of equation chosen was

$$y + \cdot a = bx^c$$

or

$$\log (y + a) = \log b + c \log x$$

where y is the rate of change of wage rates and x is the percentage of unemployment. The constants b and c were estimated by least squares using the values of y and x corresponding to the crosses in the four intervals between 0 and 5 percent unemployment, the constant a being chosen by trial and error to make the curve pass as close as possible to the remaining two crosses in the intervals between 5 and 11 percent unemployment.[4] The equation of the fitted curve is

$$y + 0.900 = 9.638x^{-1.394}$$

or

$$\log (y + 0.900) = 0.984 - 1.394 \log x$$

[4] At first sight it might appear preferable to carry out a multiple regression of y on the variables x and dx/dt. However, owing to the particular form of the relation between y and x in the present case, it is not easy to find a suitable linear multiple regression equation. An equation of the form $y + a = bx^c + k \left(\dfrac{1}{x^m} \cdot \dfrac{dx}{dt} \right)$ would probably be suitable. If so, the procedure which has been adopted for estimating the relation that would hold between y and x if dx/dt were zero is satisfactory, since it can easily be shown that $1/x^m \cdot dx/dt$ is uncorrelated with x or with any power of x provided that x is, as in this case, a trend-free variable.

Considering the wage changes in individual years in relation to the fitted curve, we see that the wage increase in 1862 (see Fig. 2) is definitely larger than can be accounted for by the level of unemployment and the rate of change of unemployment, and the wage increase in 1863 is also larger than would be expected. It seems that the 12.5-percent increase in import prices between 1861 and 1862 referred to earlier (and no doubt connected with the outbreak of the American civil war) was in fact sufficient to have a real effect on wage rates by causing cost of living increases in wages that were greater than the increases which would have resulted from employers' demand for labor and that the consequent wage-price spiral continued into 1863. On the other hand, the increases in import prices of 7.6 percent between 1899 and 1900, and again between 1909 and 1910, and the increase of 7.0 percent between 1871 and 1872 do not seem to have had any noticeable effect on wage rates. This is consistent with the hypothesis just stated about the effect of rising import prices on wage rates.

Figure 3 and Figs. 5 to 8 show a very clear relation between the rate of change of wage rates and the level and rate of change of unemployment,[5] but the relation hardly appears at all in the cycle shown in Fig. 4. The wage index of Phelps Brown and Sheila Hopkins, from which the changes in wage rates were calculated, was based on Wood's earlier index,[6] which shows the same stability during these years. From 1880 we also have Bowley's index of wage rates.[7] If the rate of change of money wage rates for 1881 to 1886 is calculated from Bowley's index by the same method as was used before, the results shown in Fig. 4a are obtained, giving the typical relation between the rate of change of wage rates and the level and rate of change of unem-

---

[5] Since the unemployed figures used are the averages of monthly percentages, the first central difference is again the best simple approximation to the average rate of change of unemployment during a year. It is obvious from an inspection of Fig. 3 and Figs. 5 to 8 that in each cycle there is a close relationship between the deviations of the points from the fitted curve and the first central differences of the employment figures, though the magnitude of the relation does not seem to have remained constant over the whole period.

[6] See Phelps Brown and Sheila Hopkins, pp. 264–65.

[7] A. L. Bowley, *Wages and Income in the United Kingdom since 1860*, Table VII, p. 30.

ployment. It seems possible that some peculiarity may have occurred in the construction of Wood's index for these years. Bowley's index for the remainder of the period up to 1913 gives results that are broadly similar to those shown in Figs. 5 to 8, but the pattern is rather less regular than that obtained with the index of Phelps Brown and Sheila Hopkins.

From Fig. 6 it can be seen that wage rates rose more slowly than usual in the upswing of business activity from 1893 to 1896 and then returned to their normal pattern of change; but with a temporary increase in unemployment during 1897. This suggests that there may have been exceptional resistance by employers to wage increases from 1894 to 1896, culminating in industrial strife in 1897. A glance at industrial history confirms this suspicion. During the 1890s there was a rapid growth of employers' federations, and from 1895 to 1897 there was resistance by the employers' federations to trade union demands for the introduction of an eight-hour working day, which would have involved a rise in hourly wage rates. This resulted in a strike by the Amalgamated Society of Engineers, countered by the Employers' Federation with a lock-out which lasted until January 1898.

From Fig. 8 it can be seen that the relation between wage changes and unemployment was again disturbed in 1912. From the monthly figures of percentage unemployment in trade unions, we find that unemployment rose from 2.8 percent in February 1912 to 11.3 percent in March, falling back to 3.6 percent in April and 2.7 percent in May, as the result of a general stoppage of work in coal mining. If an adjustment is made to eliminate the effect of the strike on unemployment, the figure for the average percentage unemployment during 1912 would be reduced by about 0.8 percent, restoring the typical pattern of the relation between the rate of change of wage rates and the level and rate of change of unemployment.

From a comparison of Figs. 2 to 8 it appears that the width of loops obtained in each trade cycle has tended to narrow, suggesting a reduction in the dependence of the rate of change of wage rates on the rate of change of unemployment. There seem to be two possible explanations of this. First, in the coal and steel industries before the First World War, sliding scale adjustments were common, by which wage rates were linked to the prices of the products. Given the tendency of product prices to rise with an increase in business activity and fall with a decrease in business activity, we see that these agreements may have strengthened the relation between changes in wage rates and changes in unemployment in these industries. During the earlier years of the period these industries would have fairly large weights in the wage index, but with the greater coverage of the statistical material available in later years the weights of these industries in the index would be reduced. Second, it is possible that the decrease in the width of the loops resulted not so much from a reduction in the dependence of wage changes on changes in unemployment as from the introduction of a time lag in the response of wage changes to changes in the level of unemployment, caused by the extension of collective bargaining and particularly by the growth of arbitration and conciliation procedures. If such a time lag existed in the later years of the period, the wage change in any year should be related, not to average unemployment during that year, but to the average unemployment lagged by, perhaps, several months. This would have the effect of moving each point in the diagrams horizontally part of the way toward the point of the preceding year, and it can easily be seen that this would widen the loops in the diagrams. This fact makes it difficult to discriminate at all closely between the effect of time lags and the effect of dependence of wage changes on the rate of change of unemployment.

3. 1913–48

A scatter diagram of the rate of change of wage rates and percentage unemployment for the years 1913–48 is shown in Fig. 9. From 1913 to 1920 the series used are a continuation of those used for the period 1861–1913. From 1921 to 1948 the Ministry of Labor's index of hourly wage rates at the end of December of each year has been used, the percentage change in the index each year being taken as a measure of the average rate of change of wage rates during that year. The Ministry of Labor's figures for the percentage unemployment in the United Kingdom have been used for the years 1921–45. For the years 1946–48 the unemployment figures were taken from the *Statisti-*

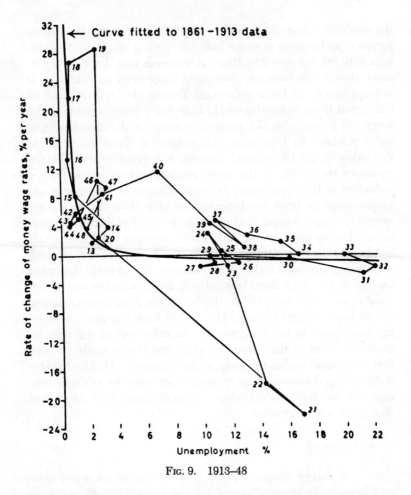

FIG. 9.   1913–48

cal Yearbooks of the International Labor Organization.

It will be seen from Fig. 9 that there was an increase in unemployment in 1914 (mainly due to a sharp rise in the three months following the commencement of the war). From 1915 to 1918 unemployment was low and wage rates rose rapidly. The cost of living was also rising rapidly and formal agreements for automatic cost of living adjustments in wage rates became widespread, but it is not clear whether the cost of living adjustments were a real factor in increasing wage rates or whether they merely replaced increases that would in any case have occurred as a result of the high demand for labor. Demobilization brought increased unemployment in 1919, but wage rates continued to rise rapidly until 1920, probably as a result of the rapidly rising import prices, which reached their peak in 1920, and consequent cost of living adjustments in wage rates. There was then a sharp increase in unemployment from 2.6 percent in 1920 to 17.0 percent in 1921, accompanied by a fall of 22.2 percent in wage rates in 1921. Part of the fall can be explained by the extremely rapid increase in unemployment, but a fall of 12.8 percent in the cost of living, largely a result of falling import prices, was no doubt also a major factor. In 1922 unemployment was 14.3 percent and wage rates fell by 19.1 percent. Although unemployment was high in this year, it was decreasing, and the major part of the large fall in wage rates must be explained by the fall of 17.5 percent in the cost of living index between 1921 and 1922. After this experience trade unions became less enthusiastic about agreements for automatic cost of living adjustments and the number of these agreements declined.

From 1923 to 1929 there were only small changes in import prices and in the cost of living. In 1923 and 1924 unemployment was high, but decreasing. Wage rates fell slightly in 1923 and rose by 3.1 percent in 1924. It seems likely that, if business activity had continued to improve after 1924, the changes in wage rates would have shown the usual pattern of the recovery phase of earlier trade cycles. However, the decision to check demand in an attempt to force the price level down in order to restore the gold standard at the prewar parity of sterling prevented the recovery of business activity, and unemployment remained fairly steady between 9.7 percent and 12.5 percent from 1925 to 1929. The average level of unemployment during these five years was 10.94 percent and the average rate of change of wage rates was − 0.60 percent per year. The rate of change of wage rates calculated from the curve fitted to the 1861–1913 data for a level of unemployment of 10.94 percent is −0.56 percent per year, in close agreement with the average observed value. Thus the evidence does not support the view, which is sometimes expressed, that the policy of forcing the price level down failed because of increased resistance to downward movements of wage rates.

The actual results obtained, given the levels of unemployment that were held, could have been predicted fairly accurately from a study of the prewar data, if anyone had felt inclined to carry out the necessary analysis.

The relation between wage changes and unemployment during the 1929–37 trade cycle follows the usual pattern of the cycles in the 1861–1913 period, except for the higher level of unemployment throughout the cycle. The increases in wage rates in 1935, 1936, and 1937 are perhaps rather larger than would be expected to result from the rate of change of employment alone, and part of the increases must probably be attributed to cost of living adjustments. The cost of living index rose 3.1 percent in 1935, 3.0 percent in 1936 and 5.2 percent in 1937, the major part of the increase in each of these years being due to the rise in the food component of the index. Only in 1937 can the rise in food prices be fully accounted for by rising import prices; in 1935 and 1936 it seems likely that the policies introduced to raise prices of home-produced agricultural produce played a significant part in increasing food prices and so the cost of living index and wage rates. The extremely uneven geographical distribution of unemployment may also have been a factor tending to increase the rapidity of wage changes during the upswing of business activity between 1934 and 1937.

Increases in import prices probably contributed to the wage increases in 1940 and 1941. The points in Fig. 9 for the remaining war years show the effectiveness of the economic controls introduced. After an increase in unemployment in 1946 due to demobilization and in 1947 due to the coal crisis, we return in 1948 almost exactly to the fitted relation between unemployment and wage changes.

4.   1948–57

A scatter diagram for the years 1948–57 is shown in Fig. 10. The unemployment percentages shown are averages of the monthly unemployment percentages in Great Britain during the calendar years indicated, taken from the *Ministry of Labor Gazette*. The Ministry of Labor does not regularly publish figures of the percentage unemployment in the United Kingdom;

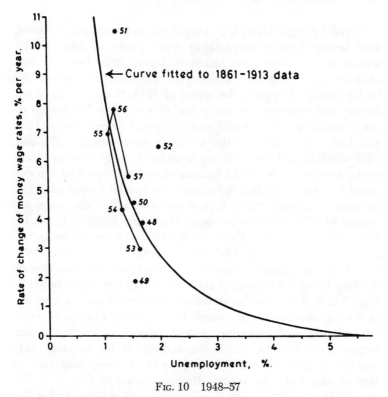

FIG. 10   1948–57

but from data published in the *Statistical Yearbooks* of the International Labor Organization it appears that unemployment in the United Kingdom was fairly consistently about 0.1 percent higher than that in Great Britain throughout this period. The wage index used was the index of weekly wage rates, published monthly in the *Ministry of Labor Gazette*, the percentage change during each calendar year being taken as a measure of the average rate of change of money wage rates during the year. The Ministry does not regularly publish an index of hourly wage rates; but an index of normal weekly hours published in the *Ministry of Labor Gazette* of September 1957 shows a reduction of 0.2 per cent in 1948 and in 1949, and an average annual reduction of approximately 0.04 percent from 1950 to 1957. The percentage changes in hourly rates would therefore be greater than the percentage changes in weekly rates by these amounts.

It will be argued later that a rapid rise in import prices during 1947 led to a sharp increase in retail prices in 1948, which tended to stimulate wage increases during 1948, but that this tendency was offset by the policy of wage restraint introduced by Sir Stafford Cripps in the spring of 1948; that wage increases during 1949 were exceptionally low as a result of the policy of wage restraint; that a rapid rise in import prices during 1950 and 1951 led to a rapid rise in retail prices during 1951 and 1952 which caused cost of living increases in wage rates in excess of the increases that would have occurred as a result of the demand for labor, but that there were no special factors of wage restraint or rapidly rising import prices to affect the wage increases in 1950 or in the five years from 1953 to 1957. It can be seen from Fig. 10 that the point for 1950 lies very close to the curve fitted to the 1861–1913 data and that the points for 1953 to 1957 lie on a narrow loop around this curve, the direction of the loop being the reverse of the direction of the loops shown in Figs. 2 to 8. A loop in this direction could result from a time lag in the adjustment of wage rates. If the rate of change of wage rates during each calendar year is related to unemployment lagged seven months, i.e. to the average of the monthly percentages of unemployment from June of the preceding year to May of that year, the scatter diagram shown in Fig. 11 is obtained. The loop has now disappeared, and the points for the years 1950 and 1953 to 1957 lie closely along a smooth curve that coincides almost exactly with the curve fitted in the 1861–1913 data.

In Table 1 the percentage changes in money wage rates during the years 1948–57 are shown in Column (1). The figures in Column (2) are the percentage changes in wage rates calculated from the curve fitted to the 1861–1913 data corresponding to the unemployment percentages shown in Fig. 11, i.e. the average percentages of unemployment lagged seven months. On the hypothesis that has been used in this paper, these figures represent the percentages by which wage rates would be expected to rise, given the level of employment for each year, as a result of employers' competitive bidding for labor, i.e. they represent the "demand pull" element in wage adjustments.

The relevant figure on the cost side in wage negotiations is the

TABLE 1

	(1) Change in Wage Rates	(2) Demand Pull	(3) Cost Push	(4) Change in Import Prices
1947				20·1
1948	3·9	3·5	7·1	10·6
1949	1·9	4·1	2·9	4·1
1950	4·6	4·4	3·0	26·5
1951	10·5	5·2	9·0	23·3
1952	6·4	4·5	9·3	−11·7
1953	3·0	3·0	3·0	−4·8
1954	4·4	4·5	1·9	5·0
1955	6·9	6·8	4·6	1·9
1956	7·9	8·0	4·9	3·8
1957	5·4	5·2	3·8	−7·3

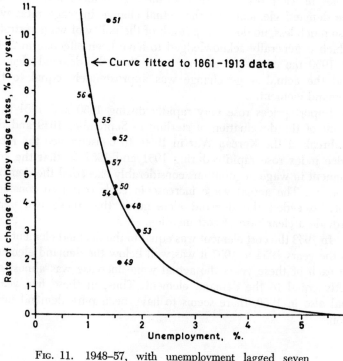

Fig. 11. 1948–57, with unemployment lagged seven months

percentage increase shown by the retail price index in the month in which the negotiations are proceeding over the index of the corresponding month of the previous year. The average of these monthly percentages for each calendar year is an approximate measure of the "cost push" element in wage adjustments, and these averages are given in Column (3). The percentage change in the index of import prices during each year is given in Column (4).

From Table 1 we see that in 1948 the cost push element was considerably greater than the demand pull element, as a result of the lagged effect on retail prices of the rapid rise in import prices during the previous year, and the change in wage rates was a little greater than could be accounted for by the demand pull element. It would probably have been considerably greater but for the cooperation of the trade unions in Sir Stafford Cripps' policy of wage restraint. In 1949 the cost element was less than the demand element and the actual change in wage rates was also much less, no doubt as a result of the policy of wage restraint which is generally acknowledged to have been effective in 1949. In 1950 the cost element was lower than the demand element and the actual wage change was approximately equal to the demand element.

Import prices rose very rapidly during 1950 and 1951 as a result of the devaluation of sterling in September 1949 and the outbreak of the Korean War in 1950. In consequence the retail price index rose rapidly during 1951 and 1952 so that the cost element in wage negotiations considerably exceeded the demand element. The actual wage increase in each year also considerably exceeded the demand element so that these two years provide a clear case of cost inflation.

In 1953 the cost element was equal to the demand element and in the years 1954 to 1957 it was well below the demand element. In each of these years the actual wage increase was almost exactly equal to the demand element. Thus, in these five years, and also in 1950, there seems to have been pure demand inflation.

5. CONCLUSIONS

The statistical evidence in Sections 2 to 4 seems in general to support the hypothesis stated in Section 1, that the rate of change of money wage rates can be explained by the level of unemployment and the rate of change of unemployment, except in or immediately after those years in which there is a sufficiently rapid rise in import prices to offset the tendency for increasing productivity to reduce the cost of living.

Ignoring years in which import prices rise rapidly enough to initiate a wage-price spiral, which seems to occur very rarely except as a result of war, and assuming an increase in productivity of 2 percent per year, it seems from the relation fitted to the data that, if aggregate demand were kept at a value that would maintain a stable level of product prices, the associated level of unemployment would be a little under 2½ percent. If, as is sometimes recommended, demand were kept at a value that would maintain stable wage rates, the associated level of unemployment would be about 5½ percent.

Because of the strong curvature of the fitted relation in the region of low percentage unemployment, there will be a lower average rate of increase of wage rates if unemployment is held constant at a given level than there will be if unemployment is allowed to fluctuate about that level.

These conclusions are, of course, tentative. There is need for much more detailed research into the relations between unemployment, wage rates, prices, and productivity.

Fluctuations in the Savings-Income Ratio: A Problem in Economic Forecasting

FRANCO MODIGLIANI

Franco Modigliani is Professor of Economics at
Massachusetts Institute of Technology. This paper
appeared in *Studies in Income and Wealth*, published by
the National Bureau of Economic Research in 1949.

1. THE PROBLEM

Considerable attention has been given in recent literature to the problem of forecasting the main components of
national income, especially the level of savings at full employment in an early post-transitional year such as 1950. Almost
invariably the method consists in projecting the relations between
national income or gross national product and its components
observed in the decades of the 'twenties and 'thirties (or only in
1929–40) to the much higher level of income expected for the
post-transitional period. The aim of this paper is to show that,
because of the violent cyclical fluctuations that characterized
the period of observation, the results tend to be systematically
biased. Some criticism on this score has already been voiced by

other authors, but no systematic attempt has apparently been made to formulate it precisely, to test its validity, and to indicate its quantitative implications for purposes of estimation and of forecasting.

We shall examine several relations among economic variables, show that there is evidence of a pronounced discrepancy between the cyclical, or short-run, and the secular, or long-run, form of these relations, and suggest methods of analysis by which it seems possible to estimate both. Although the results are tentative and leave many questions unanswered, it is hoped that the broad lines of approach suggested will be of some use in improving the reliability of our long-range, as well as short-run, forecasts.

2. RECENT ESTIMATES OF THE CONSUMPTION FUNCTION

Starting with the crucial question of forecasting savings from disposable income, a procedure for which several methods have been employed, we consider briefly those suggested by Mosak, Woytinsky, and especially Smithies.[1]

Mosak's method is most open to criticism. It consists in using the relation between consumption expenditure and disposable income in current prices observed during 1929–40 as a first approximation to the consumption function of the American economy.[2] Applying this relation to his forecast of disposable income in 1950, which, assuming 1940 rates of taxation, would amount to about 154 billion current dollars, Mosak obtains an estimate of individual savings of nearly $22 billion, some 14 percent of disposable income.

This approach has been severely criticized by Woytinsky who points out the logical difficulties involved in extrapolating a rela-

[1] J. L. Mosak, "Forecasting Postwar Demands: III," *Econometrica*, 13, No. 1 (January 1945), pp. 25–53; W. S. Woytinsky, "Relationship between Consumers' Expenditures, Savings, and Disposable Income," *Review of Economic Statistics*, 28, No. 21 (February 1946), pp. 1–12; Arthur Smithies, "Forecasting Postwar Demand: I," *Econometrica*, 13, No. 1 (January 1945), pp. 1–14.
[2] Mosak's paper indicates that he was aware of the oversimplification involved in his approach, but felt that it would not affect his results unduly.

tion between undeflated dollar series. More generally, the relation between undeflated dollar series tends to be systematically biased, especially when the time series are for a period characterized by cyclical fluctuations as violent as those of 1929–40. Theoretical considerations, as well as statistical evidence, indicate that there is a marked tendency for prices to fluctuate together with physical quantities during a cycle. The cyclical covariation of prices in turn tends to cause a marked positive correlation between the dollar series, even if the "true" relation between the series in real terms is slight or negative.

Hence the relation between series in current prices, even if more pronounced than that between the corresponding deflated series (as it often is) is an unreliable tool of analysis; extrapolation of such a relation implies, among other things, extrapolating the cyclical relation between movements of real income and prices, a particularly unjustified procedure in *long-range* forecasting.

These remarks explain in part why Mosak's formula leads to untenable results when applied to estimates of disposable income and consumption for the years immediately preceding World War I, when prices were disproportionately lower than income as compared with 1929–40. Thus, for 1913 when, according to the latest estimates of the Department of Commerce, disposable income amounted to about $33 billion and savings to $3 billion, Mosak's formula gives a level of savings of −$2 billion! If this formula fails so completely when extrapolated only 15 years back to a much lower level of income, we cannot put much confidence in the results of extrapolating it 10 years forward to a much higher level.[3]

None of the objections leveled against Mosak's type of approach can be raised against the extrapolation of the equation used by Smithies:

$$C = 76.58 + .76Y + 1.15(t - 1922) \qquad (1)$$

[3] Note that the relations between 1913 and the period of observation, on the one hand, and that between 1950 and the period of observation, on the other, are similar in many respects. Both years are separated from the period of observation by a major war with a marked price rise. Also, the level of income that may be expected to prevail in 1950 at full employment is likely to be nearly as much above the average income for the period of observation as the income prevailing in 1913 is below it.

C denotes real consumption, Y real income per capita (both in 1929 dollars), and t time. In the first place, Smithies' equation is based on the relation between deflated series (the deflator being the cost of living index); and in the second, the period of observation includes the seven relatively stable years 1923–29.

Despite these differences in approach, when Smithies' formula is applied to forecasting the level of individual savings at full employment in 1950, the results are strikingly similar to Mosak's. Thus, for the disposable income of $154 billion in 1943 prices (corresponding to Smithies' assumption C with respect to the tax structure) he, too, estimates savings to be $21–22 billion, some 14 percent of disposable income.

These results deserve closer examination. Smithies' equation not only gives a very close fit for the period of observation, but also, contrary to Mosak's, appears to explain satisfactorily the relation between income and consumption prevailing in earlier decades. In Smithies' own words, "applying the above formula to changes in Kuznets' national income figures, we obtain a close approximation to changes in his consumption figures." [4] This, in turn, raises an interesting question. As is well known, Kuznets' estimates indicate that the ratio of consumption to net national product has remained remarkably stable in the five decades 1879–88 to 1919–28, fluctuating between a minimum of 88 and a maximum of 89.2 percent, and showed no tendency to fall with the secular increase in income. Similarly, according to the Department of Commerce estimates, the average ratio of consumers' expenditures to disposable income in 1923–40 amounted to about 91 percent and remained consistently above 88 percent (except in 1923 when it was 87.4 percent). If Smithies' formula satisfactorily explains the relation between income and consumption prevailing in this period, why does application of the same for-

---

[4] Smithies, p. 6. Kuznets' figures are those given in his "Uses of National Income in Peace and War," National Bureau of Economic Research, *Occasional Paper 6*, March 1942, p. 31, Table 2; and p. 35, Table 6. These estimates were somewhat revised in Kuznets' later study, *National Income: A Summary of Findings* (National Bureau of Economic Research, 1946). Since this book was not published until after our study had been completed, the discussion in the rest of this section is based on the earlier estimates referred to by Smithies. Certain implications of the new estimates are, however, discussed in Footnote 7.

mula, when extrapolated to 1950, yield a consumption-income ratio of only 86 percent?

To answer this question we must note that, according to Smithies' equation, the consumption-income ratio depends upon the rate at which income grows. From this equation we can, in fact, derive the following:

$$\frac{C_t}{Y_t} = \frac{(76.6 + 1.15t')}{Y_t} + .76; \ (t' \text{ denotes } t - 1922) \qquad (2)$$

Equation (2) shows that the ratio C_t/Y_t will tend to rise, fall, or remain constant depending upon whether the fraction on the right side tends to rise, fall, or remain constant; and this, in turn, obviously depends upon the relation between the coefficient of t and the actual rate of growth of income in time. In particular, the income-consumption ratio will tend to fluctuate around a constant level if

$$\frac{76.6 + 1.15t'}{Y_t} + .76 = \propto$$

that is, if Y grows at the specific rate given by the formula:

$$Y_t = \frac{76.6}{\propto - .76} + \frac{1.15t'}{\propto - .76} \qquad (2a)$$

In the five decades covered by Kuznets, the consumption-income ratio fluctuated around .89. If we substitute this figure for $\propto$ in formula (2a), the coefficient of t is approximately 8.8. In other words, it appears that, if income per capita is growing in the long run at the average rate of about $8.8 per year, *then, and only then,* will the consumption-income ratio computed from Smithies' equation tend to fluctuate around a constant long-run level of .89. The average growth of income per head in Kuznets' estimates happens to be precisely $8.5 per year. This, then, explains why Smithies' formula seems consistent with the constancy of the savings-income ratio exhibited by Kuznets' estimates. On the other hand, according to Smithies' forecast, disposable income at full employment in 1950 would amount to $1060–1070 per capita in 1929 prices, while the corresponding figure in 1940 was only

$675. His forecast therefore implies an increase in real income per
capita of nearly $40 per year from 1940 to 1950. With such an
unprecedented rate of growth, Smithies' formula naturally leads
to a savings-income ratio 20 to 30 percent higher than that implicit in Kuznets' historical estimates, and 40 and 60 percent
higher than the ratio of saving to disposable income in the 'twenties and 'thirties, as estimated by the Department of Commerce.

We do not intend to discuss here whether the optimism of
Smithies and of many other investigators in forecasting such a
stupendous growth in the years to come is at all justified.[5] The
question that interests us here is whether, assuming the correctness of this forecast, we can put much confidence in the projection of Smithies' formula to a period when income is assumed to
be rising at a rate about eight times as high as during the period
of observation, and five times as high as during the period covered
by Kuznets' data, to which this formula was applied. In other
words, is the rise in the savings-income ratio (the fall in the consumption-income ratio) that follows from Smithies' formula for
periods of rapid rise in income acceptable in the light of statistical experience?

Table 1 contains a partial answer. Although Smithies' formula
gives a surprisingly good approximation to the actual total
change in savings over the period as a whole, it fails rather badly
in each subperiod in which the growth of income was markedly
different from the critical rate of $8.8 per year (Column 4).
The reason is not hard to find. As we have just seen, according to
Smithies' formula the consumption-income ratio depends upon
the rate at which income grows. Kuznets' estimates, on the other
hand, show that the fluctuations in this ratio were not only very
small but were essentially unrelated to the rate of growth of
income. It must be noted in particular that in the last decade
covered by Table 1, when per capita income rose at an annual

---

[5] It is true by 1941 income per capita had already risen to about $775.
Still, Smithies' forecast should imply an annual increase from 1941 to 1950
of some $30 per head. Inasmuch as in 1941 we were very close to full em-
ployment, this rise in income would have to be brought about almost exclusively by increases in productivity. (Also Smithies' forecast of disposable
income includes a small amount of nonproduced income or net transfer
payments; these, however, represent less than 2 percent of disposable
income.)

rate of $16, or twice as high as the critical rate, Smithies' formula is biased distinctly upward, indicating an increase in savings 26 percent larger than the actual increase. How then can we apply this formula with confidence to a period in which income is supposed to grow at an even faster rate?

Since the publication of Smithies' paper, the Department of Commerce has made available revised estimates of disposable income and consumption for 1919–28. If Smithies' method is applied to these revised data, the results are:

$$C = 71.7 + .78Y + .83(t - 1922) \qquad (3)$$

TABLE 1

CHANGES IN KUZNETS' ESTIMATES OF SAVINGS COMPARED WITH CHANGES COMPUTED BY SMITHIES' FORMULA

Period (1)	Changes in Average Annual Savings (billions of current $)		Percent of Error (3) − (2) (4)	Average Annual Rate of Change in Real Income per Capita ($, 1929 prices) (5)
	As given by Kuznets [a] (2)	As est. by Smithies' formula [b] (3)	(2)	
1879–1888 to 1889–1898	.42	−.08	−120	+3
1884–1893 to 1894–1903	.40	+.19	−52	+5.6
1889–1898 to 1899–1908	.84	+.83	−1	+8.8
1894–1903 to 1904–1913	1.17	1.08	−8	+8.1
1899–1908 to 1909–1918	1.83	1.29	−30	+6.1
1904–1913 to 1914–1923	3.57	2.83	−21	+8.7
1909–1918 to 1919–1928	3.63	4.58	+26	+16.1
1879–1888 to 1919–1928	6.72	6.62	−1	+8.5

[a] "Uses of National Income in Peace and War," National Bureau of Economic Research, *Occasional Paper 6*, 1942, p. 31, Table 2, Column 2. Kuznets uses the term "net capital formation" rather than "savings."

[b] The figures in this column were computed as follows: Kuznets' net national product (Column 1) was deflated by using the price index implicit in his conversion of consumer outlay to 1929 prices (p. 35, Table 6), and divided by the average population for each decade (*Statistical Abstract of the United States, 1944–45*, p. 8). Substituting the resulting real income per capita series in Smithies' equation and giving to t the value corresponding to the middle of each decade, we obtain a series of computed changes in real consumption per capita.

At this point there were two possible lines of procedure. We could subtract Kuznets' changes in real consumption and Smithies' changes in real consumption from Kuznets' changes in real income, thereby obtaining a true and "computed" series of changes in real savings per capita. This comparison, not given in the table since we are more interested in comparing a change in aggregate savings in current dollars than changes in real savings per capita, shows percentages of error considerably greater than those in the table, ranging

The multiple correlation coefficient, though somewhat lower than that originally obtained by Smithies, remains very high, .991.[6] What is significant, however, is the sizable fall in the coefficient of time—from 1.15 to .83. The new time trend of consumption is no longer in line with the rate of growth of Kuznets' national income; the new equation therefore gives a distinctly worse approximation even for the aggregate change in savings from the first to the last decade.[7]

To conclude: Although Smithies' hypothesis is theoretically consistent, his contention that it explains past developments satisfactorily is not fully warranted. In fact, if we accept Smithies' hypothesis that consumption depends essentially on current income, plus a trend factor entirely independent of income, we must accept also the hypothesis that the apparent long-run stability of the savings-income ratio is essentially due to chance, that is, to the coincidence of the time trend of income with the "independent" time trend of consumption. The latter hypothesis, however, is obviously not very satisfactory and, furthermore, does not stand up well under closer examination of the data.

This criticism leads us to formulate a counterhypothesis: (1) the apparent long-run stability of the savings-income ratio in the course of the gradual secular expansion of income is not due to chance, but rather to a structural property of the system, a con-

---

from a maximum of 150 percent for the first period to a minimum of 21 and 18 percent for the third and fourth periods, respectively, and amounting to 73 percent for the last period.

The other possible procedure—the one followed—was to transform the series of "computed" *changes* in "real consumption per capita" into a series of "computed" real consumption per capita, adjusting the constant term in Smithies' equation so that actual and computed series would agree in the last decade. This series was then converted into aggregate consumption at current prices by multiplying it by the price index and average decade population. Subtracting this series from Kuznets' series of net national product in current prices, we obtain "computed" average yearly savings in current prices for each decade. From this series we computed changes in savings (Column 3).

[6] Smithies has informed us that the correct figure should be .996, rather than .97 as given in "Forecasting Postwar Demand: I," p. 6, Footnote 2.

[7] Kuznets' revised estimates in *National Income: A Summary of Findings* imply an upward revision of the ratio of net capital formation to income up to the decade 1914–23 (p. 53, Table 16). While we did not recompute our Table 1 on the basis of the revised estimates (given in full in Kuznets' *National Income since 1869*, National Bureau of Economic Research, 1946), there is reason to believe that Smithies' equation, especially after the revision of the time trend indicated in the text, would significantly underestimate savings in the early decades.

sistent phenomenon that can be extrapolated; (2) the tendency for savings to fluctuate together with and approximately more than income, which according to the available evidence has been very pronounced in the interwar decades, is a cyclical phenomenon.

The hypothesis that the relation between savings, consumption, and income might be influenced by cyclical conditions has already been advanced by other authors and has recently been tested by Woytinsky. His approach, however, is not very convincing inasmuch as he segregated 1931–34 from 1923–40, and fitted separate equations to 1931–34 and to the remaining years (1923–30 and 1935–40). This procedure seems to us too arbitrary and we see no reason therefore to place much confidence in the extrapolation of the various regression equations Woytinsky obtained for the "more or less prosperous years" which indicate that the savings-income ratio tends to fall as income rises. The distinction between prosperity and depression is obviously quantitative, not qualitative, and can therefore be measured. This idea will be developed in the next section as we proceed to formulate our hypothesis more precisely, to demonstrate that it can be tested statistically, and to show that there is support for it.

3. AN ALTERNATIVE HYPOTHESIS TESTED FOR THE UNITED STATES

First we formulate operational definitions of what we mean by "cyclical" and "secular" changes in income. By the secular movement of income we mean a movement that carries real income per capita above the highest level reached in any preceding year; by cyclical movement we mean any movement, whether upward or downward, that leaves real income per capita below the highest previous peak.[8] These definitions may be conveniently given in symbolic terms. Let Y_t denote real income per capita in the year t and $Y_t{}^\circ$ denote the highest real income per capita realized in any year preceding t; the change in income between the

---

[8] This is in accordance with Marshall's use of "secular," since an expansion in income above the highest previous peak must, in general, be due to the gradual secular improvement in technology and/or an increase in capital per worker.

year t and the year $(t + 1)$ will be called cyclical, if both Y_t and $Y_{t+1} < Y_t^° = Y_{t}^°{}_{+1}$; otherwise, it will be called secular. The quantity $(Y_t - Y_t^°)/Y_t$ will be referred to as the "cyclical income index." In terms of the above definitions and symbols, the hypothesis we offer states that the proportion of income saved will be positively related to, and largely explained by, the cyclical income index.

In Chart 1 the savings-income ratio is plotted against the cyclical income index for the twenty years 1921–40, both quantities computed from the latest Department of Commerce estimates. Evidently, between the two variables there is a marked direct relation which appears to be essentially linear. The coefficient of correlation is .84, definitely significant in view of the relatively large number of observations. The regression equation is

$$S_t/Y_t = .098 + .125(Y_t - Y_t^°)/Y_t \qquad (4)$$

or

$$C_t/Y_t = .902 - .125(Y_t - Y_t^°)/Y_t$$

According to this equation,[9] if income were secularly constant, savings would be about 10 percent of income, but if income rose r percent above or fell r percent below the previous peak, the ratio would change $.125(.01)\,r$. (For instance, with a secular growth of 5 percent, the ratio would amount to approximately 10½ percent, while with a cyclical fall of 30 percent, it would amount to only about 6 percent.)

Our correlation coefficient cannot be compared directly with

[9] The analysis in the text refers exclusively to 1921–40, though the Department of Commerce has prepared estimates of disposable income and consumption as far back as 1909. It is, however, generally conceded that the margin of error in the estimates of savings for the period before 1919 is sufficiently wide to make the inclusion of these years inadvisable. Their inclusion would, we believe, reduce rather than increase the reliability of the regression equation for purposes of extrapolation. For the sake of comparison, we might add that with the inclusion of 1910–14 the correlation coefficient falls somewhat—from .84 to .77—and the regression equation changes only slightly:

$$\frac{S_t}{Y_t} = .094 + .114\,\frac{(Y_t - Y_t^°)}{Y_t}$$

Finally, if we extrapolate Eq. (4) back to 1910–14 we get, in all cases, a distinctly better approximation to the Department of Commerce estimates of consumption than by extrapolating Smithies' Eq. Both equations underestimate the consumption income ratio; in Eq. (4), the underestimate ranges from 0 to 3 percent.

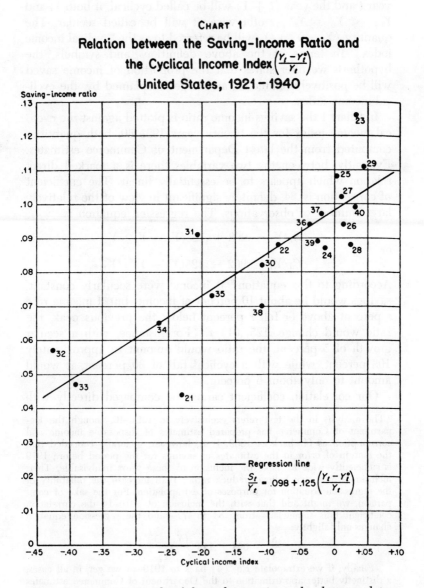

CHART 1

Relation between the Saving-Income Ratio and
the Cyclical Income Index $\left(\dfrac{Y_t - Y_t^e}{Y_t}\right)$
United States, 1921-1940

Saving-income ratio

Cyclical income index

— Regression line
$\dfrac{S_t}{Y_t} = .098 + .125\left(\dfrac{Y_t - Y_t^e}{Y_t}\right)$

that obtained on Smithies' hypothesis, since the latter relates consumption to income, while in our equation it is the consumption-income ratio that is explained. To make a comparison, we must restate our hypothesis in terms similar to those of Smithies; this can be done by multiplying both sides of Eq. (4) by Y_t:

$$C_t = kY_t - b(Y_t - Y_t^\circ) = (k - b)Y_t + b\,Y_t^\circ \tag{5}$$

This hypothesis can be tested by correlating consumption with Y_t and Y_t°. In making this test we shall also be answering one important objection that can be raised against our initial approach: namely that, by using the savings–income ratio instead of saving itself as a dependent variable, we are assuming *a priori* and without test that consumption is an homogeneous function of the independent variables. This assumption, however, should be tested by carrying out the correlation indicated by Eq. (5), then examining whether the constant term in the resulting equation is sufficiently small to be consistent with the hypothesis that its true value is approximately zero. If we make the correlation, the regression equation is

$$C_t = 2(\pm 32) + .773Y_t + .125Y_t^\circ \tag{6}$$

The corresponding multiple correlation coefficient is .992,[10] practically the same as the coefficient .993, obtained by applying Smithies' hypothesis to the revised Department of Commerce estimates for 1921–40.[11] Furthermore, the constant term in Equation (6) is evidently quite small and is statistically insignificant, being only a small fraction of its standard error, 32.

[10] The simple and partial correlation coefficients are $r_{cy} = .988$; $r_{cy}^\circ = .15$; $r_{yy}^\circ = .07$; $r_{cy.y}^\circ = .992$; and $r_{cy}^\circ{}_{.y} = .54$.   The partial correlation $r_{cy}^\circ{}_{.y}$ is not very high; nevertheless, in view of the large number of observations, it is statistically significant; in fact, by the usual test, its level of significance lies between 1 and 2 percent. Furthermore, the partial correlations are distinctly larger than the corresponding simple ones. Also, for the revised Smithies' equation, the partial correlation $r_{ct.y}$ is also only .59. If we use the ratio of the mean square successive difference to the variance of the residuals (hereinafter referred to as K) to test the randomness of the residuals in time, we get a value of 2.51. On a 5-percent level of significance, this value of K is not inconsistent with the hypothesis that the residuals are random. It is in this sense that we refer to K as insignificant in the discussion that follows. (See B. I. Hart and John von Neuman, "Tabulation of the Probabilities for the Ratio of the Mean Square Successive Difference to the Variance," *Annals of Mathematical Statistics,* 13, pp. 207–14.) The appropriateness and efficiency of this test for our purpose is open to considerable doubt. Nevertheless, it appears to be as good a test as is available.

This last result is of particular interest from our point of view. The equations obtained by Mosak and Smithies (and by most other investigators as well) contain relatively large positive constants, implying that the marginal propensity to consume is less than the average and that changes in income tend to produce less than proportional changes in consumption. On this basis both authors are led to conclude that the savings–income ratio is bound to increase whenever income rises (Mosak) or at least, whenever real income tends to rise at a sufficiently high average annual rate (Smithies).

Our results indicate instead that we must distinguish between (a) the short-run or cyclical marginal propensity to save, and (b) the long-run average and marginal propensity.

1. As long as income rises secularly, Y_t and $Y_t°$ will rise together. Therefore, the savings–income ratio will depend, not on income, but essentially on the rate of change in income. This can best be seen if we rewrite Eq. (6) using the identity $S_t = Y_t - C_t$:

$$S_t = -2 + .102Y_t + .125(Y_t - Y_t°)$$

Since the constant term is entirely negligible in comparison with the relevant values of Y, savings tend to represent approximately 10 percent of income plus about 12 percent of the increment of income. Because of the last term, the proportion of income saved will tend to vary somewhat in years of secular expansion, increasing as the rate of change in income accelerates.[12] But since the

---

[11] The addition of the years 1921 and 1922 to the period originally used by Smithies (1933–40) raises Smithies' multiple correlation from .991 to .993.

[12] Such a lag seems to explain, for instance, the high savings-income ratio for 1923 and 1929. It undoubtedly explains also, at least in part, the exceptionally high savings-income ratio for 1941, when, according to the Department of Commerce estimates, real income per capita increased 15 percent—three to four times more than the largest annual growth in the entire period of observation. If we extrapolate our equation to this year, we obtain a savings-income ratio that is higher than in any other year but still falls considerably short of the Department of Commerce figure, 15.9 percent. It is not unlikely that, for exceptionally high secular rates of increase above the highest previous peak, the lag of consumption may be more pronounced (and possibly last longer) than indicated by our equation; this point will be briefly considered later. Extrapolation of Smithies' equation also fails to explain the behavior of consumption in 1941, since it gives the very same figure as Eqs. (4) and (6).

normal secular growth is in the order of 2 to 3 percent, we may conclude that the savings–income ratio will tend to fluctuate around a level of about 10½ percent. This figure also clearly measures the proportion of any secular increment in income that will tend to be saved in the long run (that is, the long-run marginal propensity to save).

2. In the case of cyclical fluctuations in income, on the other hand, $Y_t°$ is fixed by definition. Hence the relation between savings and income takes the form: $S_t = -(2 + .125Y_t°) + .23Y_t$. The cyclical marginal propensity to save is given by the coefficient of Y_t or .23, as compared with the secular marginal propensity of .10 to .11. Also, on account of the constant term, the savings–income ratio tends to fluctuate with income during each cycle, falling below the secular level as income declines and rising toward it again as income increases.

SELECTED READINGS

Arrow, K., H. Chenery, B. Minhas, and R. Solow, "Capital-Labor Substitution and Economic Efficiency," *Review of Economics and Statistics,* August 1961.

Bass, F., "Marketing Research Expenditures: A Decision Model," *Journal of Business,* January 1963.

Cochran, W., *Sampling Techniques,* John Wiley, 1963.

Cochran, W., F. Mosteller, and S. Tukey, *Statistical Problems of the Kinsey Report,* American Statistical Association, 1954.

David F., *Games, Gods, and Gambling,* Hafner, 1962.

Dean, J., *The Relation of Cost to Output for a Leather Belt Shop,* National Bureau of Economic Research, 1941.

Douglas, P., "Are There Laws of Production?" *American Economic Review,* 1948.

Duesenberry, J., G. Fromm, L. Klein, and E. Kuh, *The Brookings Quarterly Econometric Model of the United States,* Rand McNally, 1965.

Duncan, A., *Quality Control and Industrial Statistics,* Richard D. Irwin, 1959.

Feller, W., *An Introduction to Probability Theory and its Applications,* John Wiley, 1950.

Fisher, R. A., *Design of Experiment,* Oliver and Boyd, 1949.

Friedman, M., *A Theory of the Consumption Function,* Princeton University Press, 1957.

Griliches, Z., *Hedonic Price Indexes for Automobiles,* Hearings before Joint Economic Committee of U.S. Congress, January 24, 1961.

Hansen, M., W. Hurwitz, and W. Madow, *Sample Survey Methods and Theory,* John Wiley, 1953.

Hauser, P., and W. Leonard, *Government Statistics for Business Use,* John Wiley, 1956.

Heady, E., "An Econometric Investigation of the Technology of Agricultural Production Functions," *Econometrica,* April 1957.

Huff, D., *How to Lie with Statistics,* W. W. Norton, 1954.

Johnston, J., *Econometric Methods,* McGraw-Hill, 1963.

Klein, L., *An Introduction to Econometrics,* Prentice-Hall, 1962.

Kruskal, W., and L. Telser, "Food Prices and the Bureau of Labor Statistics," *Journal of Business,* July 1960.

Luce, R.D., and H. Raiffa, *Games and Decisions*, John Wiley, 1957.

Mansfield, E., and H. Wein, "A Regression Control Chart for Costs," *Applied Statistics*, March 1958.

Morgenstern, O., *On the Accuracy of Economic Observations*, Princeton University Press, 1963.

National Bureau of Economic Research Price Statistics Review Committee, *The Price Statistics of the Federal Government*, National Bureau of Economic Research, 1961.

Nerlove, M., "Returns to Scale in Electricity Supply," *Measurement in Economics* (edited by C. Christ), Stanford University Press, 1963.

Orcutt, G., "Microanalytic Models of the United States Economy," *American Economic Review*, May 1962.

Pearson, K., *The Grammar of Science*, John Dent, 1949.

Raiffa, H., *Decision Analysis*, Addison Wesley, 1968.

Roberts, H., "The New Business Statistics," *Journal of Business*, January, 1960.

Savage, L., *The Foundations of Statistics*, John Wiley, 1954.

Schlaifer, R., *Probability and Statistics for Business Decisions*, McGraw-Hill, 1959.

Schultz, H., *Theory and Measurement of Demand*, University of Chicago Press, 1938.

Suits, D., "Forecasting and Analysis with an Econometric Model," *American Economic Review*, March 1962.

Tippett, L., "A Guide to Acceptance Sampling," *Applied Statistics*, November 1958.

Von Mises, R., *Probability, Statistics, and Truth*, 1939.

Wallis, W.A., "Economic Statistics and Economic Policy," *Journal of the American Statistical Association*, March 1966.

Yates, F., *Sampling Methods for Censuses and Surveys*, Griffin, 1960.